Elementary Geometry for College Students

FIFTH EDITION

Dan Alexander
Parkland College

Geralyn Koeberlein
Mahomet-Seymour High School

Prepared by

Geralyn Koeberlein
Mahomet-Seymour High School

Dan Alexander
Parkland College

BROOKS/COLE
CENGAGE Learning™

Australia • Brazil • Japan • Korea • Mexico • Singapore • Spain • United Kingdom • United States

ISBN-13: 978-1-4390-4793-4
ISBN-10: 1-4390-4793-6

Brooks/Cole
20 Davis Drive
Belmont, CA 94002-3098
USA

Cengage Learning is a leading provider of customized learning solutions with office locations around the globe, including Singapore, the United Kingdom, Australia, Mexico, Brazil, and Japan. Locate your local office at: **www.cengage.com/global**

Cengage Learning products are represented in Canada by Nelson Education, Ltd.

To learn more about Brooks/Cole, visit **www.cengage.com/brookscole**

Purchase any of our products at your local college store or at our preferred online store **www.ichapters.com**

Printed in the United States of America
1 2 3 4 5 6 7 13 12 11 10 09

<u>Contents</u>

How to Study Geometry

Textbook:

1. Read the textbook word-for-word. Give added attention to new vocabulary terms.
2. For new vocabulary terms:
   a. Make a drawing/create an example that illustrates the new concept.
   b. State the definition of the new term in your own words.
3. When you encounter a new postulate or theorem:
   a. Read/reread the statement until you understand the claim made.
   b. Make a drawing that illustrates the newfound relationship.
   c. State the postulate/theorem in your own words.

NOTE: The student may wish to create an index card containing the statement of the new postulate or theorem. Include a drawing and example that illustrate this principle.

4. When you study an example:
   a. Read the example step-by-step, justifying each step as the example unfolds.
   b. Refer to drawings that provide visual support for the steps of the example.
   c. With the textbook closed, repeat the steps of the example.
5. When you encounter a completed proof:
   a. Note the order of statements in that one conclusion leads to another.
   b. Compare each reason with the statement that it justifies.
   c. Refer to drawings as you compare statements and reasons of the proof.
   d. Compare statements in reverse order. Note that the final statement must be the "Prove" statement. Ask "What must the preceding statement of the proof have been . . . in that it would lead to the current statement?"
   e. With the textbook closed, try to repeat the steps of the proof.
6. Classify by groups; for instance, consider types of triangles or quadrilaterals. Also, list methods for proving "lines parallel" or "triangles congruent."

Assignments:

1. Complete as many of the assigned problems as possible. Do as much as possible on your own. Use the textbook for reference to find helpful terms and principles.
2. You may eventually need to work with a study buddy, a mathematics lab assistant, or visit your instructor during an office hour.
3. Use the Student Study Guide when you are stuck. Try to use a step from the solution to allow yourself to continue the solution on your own.

NOTE: Do NOT simply read the solution in the Student Study Guide; this approach will NOT help you to generate the solution in later assigned or tested problems.

4. When an assigned problem seems too difficult to manage, go on to the next problem. In some cases, you will be able to return to and complete these earlier problems.

Preparing for a Test:

1. For the test material, be familiar with the list of important concepts found in the textbook. Use the textbook or its Glossary to understand any new concept.
2. Study Chapter Summaries, solve Chapter Review Exercises, and complete Chapter Tests for the relevant chapters.
3. Form a study group that you can work with before each test.

Attendance:

1. Attend all classes! Arrive on time! Does your school have textbook related videos?
2. Stay current/keep informed . . . so that you do not lose interest by falling behind.

Section-by-Section Objectives

The following objectives are student goals for the study of geometry with *Elementary Geometry for College Students, the Fifth Edition* by Alexander and Koeberlein.

Chapter 1: Line and Angle Relationships

Section 1.1: Statements and Reasoning
1. Determine whether a collection of words/symbols forms a statement
2. Form the negation of a given statement
3. Form the conjunction, disjunction, or implication of two simple statements
4. Recognize the hypothesis/conclusion of a conditional statement (implication)
5. State and recognize the three types of reasoning (intuition, induction, or deduction) used in the study of geometry
6. Recognize/apply the Law of Detachment
7. Form the union/intersection of two given sets
8. Illustrate a deductive argument visually by using a Venn Diagram

Section 1.2: Informal Geometry and Measurement
1. Describe the terms point, line, and plane
2. Become familiar with geometric terms such as collinear, line segment, and angle
3. Measure a line segment with a ruler/an angle with a protractor
4. Write equations based upon statements involving concepts such as midpoint, etc.
5. Recognize terms such as right angle, straight angle, and perpendicular lines
6. Use the compass/straightedge to construct a line segment of given length
7. Use the compass/straightedge to construct the midpoint of a line segment

Section 1.3: Early Definitions and Postulates
1. State the four parts of a mathematical system: undefined terms, definition (defined terms), postulates, and theorems
2. Recognize the necessity of/characteristics for a precise definition
3. Use symbols to express a line, ray, line segment, and length of line segment
4. State the initial postulates for lines and planes (in your own words)
5. Apply the Segment-Addition Postulate
6. Recognize and apply the concepts parallel lines and parallel planes

Section 1.4: Angles and Their Relationships
1. Write symbols that name an angle or its measure
2. Understand the terminology for angles (like sides, vertex, etc.)
3. State/apply postulates involving an angle or angles
4. Given an angle's measure, classify it as acute, right, obtuse, straight, or reflex
5. Apply the Angle-Addition Postulate
6. Classify pairs of angles as adjacent, congruent, complementary, etc.
7. Use the compass/straightedge to construct an angle congruent to a given angle
8. Use the compass/straightedge to construct the bisector of a given angle

Section 1.5: Introduction to Geometric Proof
1. Demonstrate the two-column form of a proof
2. Understand the role of the Given, Prove, and Drawing for a proof problem
3. Provide reasons that justify statements found in proofs
4. Provide statements that are justified by principles (the reasons) found in proof

Section 1.6: Relationships – Perpendicular Lines
1. Know/apply the definition of perpendicular lines
2. Recognize/understand the general notion of *relation*
3. Understand/apply reflexive, symmetric, and transitive properties (for congruence)
4. At a point on a given line, construct the line perpendicular to the given line
5. Construct the perpendicular-bisector of a given line segment

Section 1.7: The Formal Proof of a Theorem
1. Determine the hypothesis and conclusion of a stated theorem
2. State the five written parts of the formal proof of a theorem
3. Create a "Drawing" based upon the hypothesis of a stated theorem
4. Write the "Given" based upon the "Drawing" for the proof of a stated theorem
5. Write the "Prove" based upon the "Drawing" for the proof of a stated theorem
6. State/apply theorems involving perpendicular lines, complementary angles, etc
7. Construct/complete the formal proof of a theorem

Chapter 2: Parallel Lines

Section 2.1: The Parallel Postulate and Special Angles
1. From a point not on a given line, construct the line perpendicular to that line
2. Recognize when two lines, a line and a plane, or two planes are perpendicular
3. Recognize when two lines, a line and a plane, or two planes are parallel
4. Define parallel lines and parallel planes
5. Understand/apply terms such as transversal, corresponding angles, etc
6. State/apply the initial postulates and theorems involving parallel lines

Section 2.2: Indirect Proof
1. Know the true/false relationships between a conditional statement and its converse, inverse, and contrapositive
2. State/apply the Law of Negative Inference
3. State/apply the method of indirect proof
4. Know that negations/uniqueness theorems are often proved by the indirect method

Section 2.3: Proving Lines Parallel
1. State/apply/prove selected theorems that establish that lines are parallel
2. From a point not on a given line, construct the line parallel to the given line

Section 2.4: The Angles of a Triangle
1. Know the definition of triangle and its related terms (vertices, sides, etc)

2. Classify triangles by side relationships (scalene, isosceles, equilateral)
3. Classify triangles by angle relationships (acute, right, obtuse, and equiangular)
4. Apply the theorem, "The sum of the angles of a triangle is 180°."
5. State/apply/prove the corollaries of the theorem found in (4).

Section 2.5: Convex Polygons
1. Know the definition of polygon and also its related terms
2. Classify polygons as concave or convex *and* by the number of sides
3. Determine the number of diagonals for a polygon of *n* sides
4. State/apply theorems involving sums of angle measures of a polygon
5. Classify polygons as equiangular, equilateral, or regular
6. Recognize/classify a figure that is a polygram/regular polygram

Section 2.6: Symmetry and Transformations
1. Recognize whether a plane figure has one or more lines of symmetry
2. Classify line symmetry as vertical symmetry, horizontal symmetry, or neither
3. Determine whether a plane figure has point symmetry
4. Recognize/classify/draw transformations that are slides, reflections, or rotations
5. Know that the transformation of a given figure produces a congruent figure

Chapter 3: Triangles

Section 3.1: Congruent Triangles
1. State the definition of congruent triangles.
2. Determine the correspondence between the parts of two congruent triangles
3. Name the included side for two named angles of a triangle
4. Name the included angle for two named sides of a triangle
5. Know/determine/apply the methods (SSS, SAS, ASA, and AAS) for proving that triangles are congruent

Section 3.2: Corresponding Parts of Congruent Triangles
1. Use CPCTC to abbreviate the statement "Corresponding parts of congruent triangles are congruent."
2. Recognize the types of conclusions that can be established by applying CPCTC
3. Use like markings to indicate corresponding parts of congruent triangles
4. State/apply the HL Theorem
5. State/apply the Pythagorean Theorem
6. Determine which method establishes that a certain pair of triangles are congruent

Section 3.3: Isosceles Triangles
1. In a triangle, be able to distinguish between the angle-bisector of an angle, an altitude to a side, the perpendicular-bisector of a side, and a median
2. Know that a triangle has three angle-bisectors, three altitudes, three medians, and three perpendicular-bisectors of its sides
3. Decide whether the description of a line that is to be used as an auxiliary line is determined, overdetermined, or underdetermined

4. State/apply "If two sides of a triangle are congruent, then the angles opposite these sides are congruent" and also its *converse*
5. State/apply the definition of the perimeter of a triangle

Section 3.4: Basic Constructions Justified
1. Construct/justify the construction of an angle congruent to a given angle
2. Construct/justify the angle-bisection method
3. Construct/validate a line segment of specified length
4. Construct/validate an angle of a specified measure
5. Construct/validate a regular polygon of a specified number of sides

Section 3.5: Inequalities in a Triangle
1. Know/apply the definition of "is less than"
2. State/apply lemmas and theorems involving inequalities in a triangle
3. State/apply corollaries involving the length of a line segment from a point not on a line (or plane) that is perpendicular to that line (or plane)
4. State/apply the Triangle Inequality (or its alternate form)

Chapter 4: Quadrilaterals

Section 4.1: Properties of a Parallelogram
1. State definitions of quadrilateral and parallelogram
2. State/apply/prove selected theorems involving parallelograms
3. Use angle measures of a parallelogram to determine its longer/shorter diagonal
4. Determine speed/direction of an airplane flying subject to wind conditions

Section 4.2: The Parallelogram and Kite
1. Know that the parallelogram/kite each have two pairs of congruent sides
2. Know that quadrilaterals with congruent opposite sides are parallelograms
3. Know that quadrilaterals with a pair of congruent and parallel sides are parallelograms
4. Know that quadrilaterals with diagonals that bisect each other are parallelograms
5. Know that a kite has one pair of opposite angles that are congruent
6. Know that one diagonal of a kite is the perpendicular-bisector of the other diagonal
7. State/apply the theorem in which the midpoints of two sides of a triangle are joined

Section 4.3: The Rectangle, Square, and Rhombus
1. State definitions of the rectangle, square, and rhombus
2. State/apply/prove theorems involving the rectangle/square/rhombus
3. State/apply/prove corollaries involving the rectangle/square/rhombus
4. Apply the Pythagorean Theorem with quadrilaterals

Section 4.4: The Trapezoid
1. Know the definition of trapezoid and its related terminology (bases, etc)
2. State/apply/prove theorems and corollaries involving trapezoids
3. State/apply theorems and corollaries involving isosceles trapezoids

Chapter 5: Similar Triangles

Section 5.1: Ratios, Rates, and Proportions
1. State/apply the terms ratio, rate, and proportion
2. Know the terminology (means, geometric mean, etc) related to proportions
3. State/apply the Means-Extremes Property (of a proportion)
4. Understand/apply further properties of proportions

Section 5.2: Similar Polygons and Triangles
1. Form an intuitive understanding of the concept "similarity of figures"
2. Determine the correspondences between the parts of similar polygons
3. Use correspondences of parts for similar polygons to find angle measures and lengths of sides for these polygons

Section 5.3: Proving Triangles Similar
1. Name/cite the methods (AA, SAS ~ , and SSS ~) for proving triangles similar
2. Use CASTC to prove that corresponding angles of similar triangles are congruent
3. Use CSSTP to prove that the lengths of corresponding sides of similar triangles are proportional

Section 5.4: The Pythagorean Theorem
1. State/apply/prove theorems that lead to the Pythagorean Theorem
2. State/apply the Pythagorean Theorem and its converse
3. Determine whether the triple (a,b,c) is a Pythagorean Triple
4. Use the lengths of sides of a triangle to classify the triangle as acute, right, or obtuse

Section 5.5: Special Right Triangles
1. State/apply/prove the 45-45-90 Theorem
2. State/apply/prove the 30-60-90 Theorem
3. Recognize/apply further theorems involving special triangles

Section 5.6: Segments Divided Proportionally
1. Form an intuitive understanding of the concept "segments divided proportionally"
2. State/apply the definition of segments divided proportionally
3. State/apply/prove the theorem that establishes proportional segments on transversals between parallel lines
4. State/apply the Angle-Bisector Theorem
5. Apply Ceva's Theorem to form a proportion among the parts of sides of a triangle

Chapter 6: Circles

Section 6.1: Circles and Related Segments and Angles
1. Become familiar with the terminology (radius, center, chord, etc) of the circle
2. State/apply postulates involving circle relationships
3. State/apply/prove selected theorems involving circle relationships
4. State/apply methods for measuring central and inscribed angles of a circle

Section 6.2: More Angle Measures in the Circle
1. State definitions for terms such as tangent and secant (of a circle)
2. Recognize when polygons are inscribed in/circumscribed about a circle
3. Recognize when circles are inscribed in/circumscribed about a polygon
4. State/apply methods of measuring angles with vertex inside, on, or outside a circle

Section 6.3: Line and Segment Relationships in the Circle
1. State/apply/prove theorems relating radii and chords of a circle
2. Recognize/apply terminology involving tangent circles
3. Recognize/apply terminology involving common tangents for two circles
4. State/apply theorems involving lenghs of chords, tangents, and secants

Section 6.4: Some Constructions and Inequalities for the Circle
1. Apply "The radius drawn to the point of tangency is perpendicular to the tangent."
2. Perform the construction of a tangent to a circle at a point on the circle
3. Perform the construction of a tangent to a circle from a point outside the circle
4. State/apply/prove theorems relating unequal chords, unequal arcs, etc

Chapter 7: Locus and Concurrence

Section 7.1: Locus of Points
1. State the definition of the term "locus of points"
2. Draw/describe the locus of points in a plane or in space
3. Know "The locus of points equidistant from the sides of an angle is the angle-bisector."
4. Know "The locus of points equidistant from two points is the perpendicular-bisector of the line segment having the two point for endpoints."
5. Verify a locus theorem by establishing two conditions

Section 7.2: Concurrence of Lines
1. State the definition of the term "concurrence of lines"
2. State/apply/prove the theorem verifying the concurrence of the three perpendicular-bisectors of the sides of a triangle
3. State/apply/prove the theorem verifying the concurrence of the three angle-bisectors of a triangle
4. State/apply the theorem regarding the concurrence of the three altitudes of a triangle
5. State/apply the theorem regarding the concurrence of the three medians of a triangle
6. Know these terms: incenter, circumcenter, orthocenter, and centroid of a triangle

Section 7.3: More About Regular Polygons
1. Determine whether a given polygon can be inscribed in/circumscribed about a circle
2. Perform constructions leading to inscribed/circumscribed polygons and circles
3. Know these terms related to regular polygons: radius, apothem, and central angle
4. Calculate measures of central angles, the length of radius, or the length of an apothem for specified regular polygons

Chapter 8: Areas of Polygons and Circles

Section 8.1: Area and Initial Postulates
1. Develop an intuitive understanding of the concept "area of a plane region"
2. Distinguish between units for length and units of area measurement'
3. State/apply the initial postulates involving the areas of plane regions
4. Prove/apply theorems involving the area of a square, parallelogram, or triangle

Section 8.2: Perimeter and Area of Polygons
1. State/apply perimeter formulas for selected polygons
2. State/apply Heron's Formula for the area of a triangle
3. State/apply/prove formulas for the areas of trapezoid, rhombus, and kite
4. Use the square of the ratio between the lengths of two corresponding sides of similar polygons to determine the ratio between their areas

Section 8.3: Regular Polygons and Area
1. Review the terms that relate to regular polygons
2. Know/apply the formulas for the area of a square and for an equilateral triangle
3. Determine the area of select regular polygons by applying the formula $A = \frac{1}{2}aP$

Section 8.4: The Circumference and Area of a Circle
1. Know that π is the ratio between circumference and length of diameter of a circle
2. Memorize common approximations of π such as 3.14
3. Know/apply the formulas $C = \pi d$ and $C = 2\pi r$ for the circumference of a circle
4. Understand/apply the formula for the length of arc of a circle
5. State/apply the formula $A = \pi r^2$ for the area of a circle

Section 8.5: more About Area Relationships in the Circle
1. Be able to distinguish between sector of a circle and segment of a circle
2. Understand/apply the formula for the area of a sector of a circle
3. Determine the area of a specified segment of a circle
4. For a triangle inscribed in a circle of radius r, prove that the area of a triangle (with perimeter P) is given by $A = rP$.
5. Apply the formula $A = rP$ to find the area of a triangle inscribed in a circle

Chapter 9: Surfaces and Solids

Section 9.1: Prisms, Area, and Volume
1. Understand intuitively the concept of prism
2. Understand/use terminology (edges, vertices, faces, etc) related to prisms
3. Determine the lateral area/total area of a prism
4. Understand/memorize/apply the formula for the volume of a prism $V = Bh$

Section 9.2: Pyramids, Area, and Volume
1. Understand intuitively the concept of pyramid
2. Understand/use terminology (edges, vertices, faces, etc) related to pyramids

3. Apply $\ell^2 = a^2 + h^2$, which relates the lengths of the slant height, apothem, and altitude of a regular pyramid
4. Apply $e^2 = h^2 + r^2$, which relates the lengths of the lateral edge, altitude, and radius of a regular pyramid
5. Determine the lateral area/total area of a pyramid
6. Memorize/apply the formula for the volume of a pyramid $V = \frac{1}{3}Bh$

Section 9.3: Cylinders and Cones
1. Understand intuitively the concepts of cylinder and cone
2. Understand/use terminology related to cylinders and cones
3. Apply $\ell^2 = r^2 + h^2$, which relates the lengths of the slant height, radius of base, and altitude of a right circular cone
4. Memorize/apply formulas for lateral area/total area of a right circular cylinder
5. Memorize/apply formulas for lateral area/total area of a right circular cone
6. Memorize/apply formulas for the volume of a right circular cylinder/cone

Section 9.4: Polyhedrons and Spheres
1. Understand intuitively the notion of polyhedron
2. Know/use terminology related to a polyhedron/regular polyhedron
3. Verify Euler's Formula, $V + F = E + 2$, relating the numbers of vertices, faces, and edges of a polyhedron
4. State/describe the five regular polyhedrons (tetrahedron, hexahedron, etc)
5. Know/apply the term sphere and related terminology
6. Memorize/apply formulas for the surface area/volume of a sphere
7. Understand/apply the concept "solid of revolution"

Chapter 10: Analytic Geometry

Section 10.1: The Rectangular Coordinate System
1. Know/use terms related to the rectangular coordinate system
2. Plot/read points in the rectangular coordinate system as ordered pairs
3. Find the distance between two points on a horizontal/vertical line segment
4. Know/apply/prove the Distance Formula
5. Know/apply the Midpoint Formula

Section 10.2: Graphs of Linear Equations and Slope
1. State/apply the definition of graph of an equation
2. Determine/use intercepts to graph linear equations (straight lines)
3. Know/apply the Slope Formula
4. Recognize by sight whether the slope of a given line is positive, negative, zero, or undefined
5. Use slope relationships to determine whether lines are parallel/perpendicular

Section 10.3: Preparing to do Analytic Proof
1. Determine which analytic formula is needed to prove a given statement
2. Prepare a drawing that is suitable for completing the analytic proof of a theorem

3. State general coordinates of vertices for a particular type of geometric figure
4. Use algebraic relationships to develop geometric relationships in a proof

Section 10.4: Analytic Proofs
1. Develop a logical and orderly plan for completion of an analytic proof
2. Construct the analytic proof of a given geometric theorem

Section 10.5: Equations of Lines
1. Use the Slope-Intercept and Point-Slope Forms to find equations of lines
2. Use the given equation of a line to draw its graph
3. Use graphs/algebra to solve systems of equations (find points of intersection)
4. Develop analytic proofs for theorems by using equations of lines

Chapter 11: Introduction to Trigonometry

Section 11.1: The Sine Ratio and Applications
1. Define/apply the sine ratio of an acute angle of a right triangle
2. Use a table/calculator to determine the sine ratio of an acute angle
3. Use a table/calculator to find the measure of an acute angle with known sine ratio
4. Understand/apply the notion of angle of elevation/depression

Section 11.2: The Cosine Ratio and Applications
1. Define/apply the cosine ratio of an acute angle of a right triangle
2. Use a table/calculator to determine the cosine ratio of an acute angle
3. Use a calculator to determine the measure of an acute angle with known cosine ratio
4. Know/apply/prove the trigonometric identity $\sin^2 \theta + \cos^2 \theta = 1$

Section 11.3: The Tangent Ratio and Other Ratios
1. Define/apply the tangent ratio of an acute angle of a right triangle
2. Recognize which trigonometric ratio (sine, cosine, or tangent) can be used to find an unknown measure of some part of a right triangle
3. Use a table/calculator to determine the tangent ratio of an acute angle
4. Use a calculator to find the measure of the acute angle with known tangent ratio
5. State/apply the definitions of cotangent, secant, and cosecant for an acute angle of a right triangle
6. Define/determine $\cot \theta$, $\sec \theta$, and $\csc \theta$ as reciprocals of $\tan \theta$, $\cos \theta$, and $\sin \theta$ respectively

Section 11.4: More Trigonometric Relationships
1. State/apply the formula for the area of a triangle $A = \dfrac{1}{2} ab \sin \gamma$ (or equivalent)
2. State/apply the Law of Sines
3. State/apply the Law of Cosines
4. Use the given measures for parts of a triangle to determine whether the Law of Sines/Cosines should be applied toward finding the measure of a remaining part

To the Student:

This interactive companion was designed to serve many purposes, including:
  Reinforce concepts found in the textbook;
  Encourage students to learn cooperatively;
  Have students work with instructors;
  Repeat terminology, symbols, principles, and techniques found in this textbook.

Improvements for the Interactive Companion for the Fifth Edition include additional questions, improved questions, increased accuracy, and provision of selected answers found at the end of the interactive companion.

Following a class session in which you have utilized the Interactive Companion portion of the Student Study Guide, you (the student) will be better prepared to delve more deeply into the subject matter of Geometry. After each class session, you should read each section thoroughly; in this way, you will both reinforce many concepts discovered in class while discovering other concepts that were not fully discussed during class. Following the class development and the reading of the textbook, you will be much better prepared to deal with the problems assigned for homework. Many homework problems will challenge the student in that they are theoretical or applied or provide an added level of difficulty over what was managed in the classroom session.

The Interactive Companion was never intended as an alternative to the textbook or to doing homework. When the Interactive Companion is properly used in conjunction with the textbook, the student's knowledge, skill level, and level of proficiency will naturally be increased.

Daniel C. Alexander
Geralyn M. Koeberlein

Statements

1. Complete: A _____ is a group of words and/or symbols that can collectively be classified as true or false.

2. Which of the following is/are statements? _____
   a. 3 + 4 = 12   b. 3 × 4 = 12   c. Are you leaving?
   d. Look out!   e. Billy Bob and his brother

3. For the simple statements P and Q, write their:
   a. Conjunction _____ b. Disjunction _____ .

4. Classify these statements as true or false.
   a. Babe Ruth played baseball and 3 + 4 = 12. _____ .
   b. Babe Ruth played baseball and 3 × 4 = 12. _____ .
   c. Babe Ruth played baseball or 3 + 4 = 12. _____ .
   d. Babe Ruth played baseball or 3 × 4 = 12. _____ .

5. State the negation of the given statement.
   a. Mary is a seamstress. _____ .
   b. Emma does not live in Charleston. _____ .

6. Given that statement P is true while statement Q is false, classify the following statements as true or false.
   a. not P _____   b. not Q _____   c. P and not Q _____   d. not P or Q. _____ .

7. Classify each implication (conditional statement) as true or false.
   a. If x is an even integer, then x can be divided exactly by 2. _____ .
   b. If Antonio lives in Dallas, then he lives in Texas. _____ .

Intuition, Induction, and Deduction

8. In the following statement, what does intuition suggest that you may conclude regarding the word *midpoint*? Also, see the figure.

   M is the midpoint of line segment AB. $\cdot A$ M $\cdot B$

   _____ .

9. In the following situation, what conclusion might you draw based upon does induction?
   It rained Sunday, Monday, Tuesday, and Wednesday. What do you expect of Thursday's weather?

   _____ .

Induction, intuition, and Deduction

10. In the following situation, what does deduction allow you to conclude? Assume that statements (i) and (ii) are both true.
    i. If a person grew up in Minnesota, then he/she has seen snow.
    ii. Katrina was raised in Rochester, Minnesota.
    _____.

11. In the following situation, what does deduction allow you to conclude? Assume that statements (i) and (ii) are both true.
    i. If a person lives in Acapulco, then he/she lives in Mexico.
    ii. Juan lives in Mexico.
    _____.

12. In the following situation, what does deduction allow you to conclude? Assume that statements (i) and (ii) are both true.
    i. All professional basketball players are tall.
    ii. MJ is a professional basketball player.
    _____.

Venn Diagrams

13. Let P represent the set "members of the state house of representatives" while Q represents the set "legislators with 2 year terms of office." Which Venn diagram correctly depicts the implication, "If a person is a member of the state house of representatives, then he/she is a legislator with a 2 year term of office."
    The correct diagram is the one on the _____ (left or right).

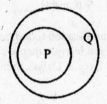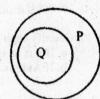

14. Use a Venn diagram to draw a conclusion, if one is possible, for the following situation. Assume that statements (i) and (ii) are both true.
    i. If an object is a skeegle, then it can be used to store water.
    ii. A bonf is a type of skeegle.
    _____.

15. Use a Venn diagram to draw a conclusion, if one is possible, regarding the angles of triangle ABC. Assume that statements (i), (ii), and (iii) are all true.
    i. If a figure is an isosceles triangle, then the triangle has 2 sides of the same length.
    ii. If a triangle has 2 sides of the same length, then it has 2 angles with the same measure.
    iii. Triangle ABC is an isosceles triangle.
    _____.

Undefined Terms

1. The three most basic undefined terms of geometry are point, _____ , and plane.

2. The notation A-X-B indicates that the points, A, X, and B are _____ and that point
   __ lies between the remaining points.

Symbols

3. The symbol ∠ represents the word _____ while the symbol for the word triangle is _____.

4. Given the line segment $\overline{CD}$, the length of this line segment is represented by _____.

Measuring Line Segments

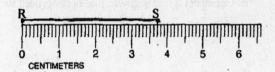

5. With the line segment $\overline{RS}$ as shown, its length is
   ___ centimeters.

6. For the line segment shown in Exercise 5, what is the distance from the midpoint
   of $\overline{RS}$ to either point R or point S? _____.

7. In the figure at right, suppose that AB = 5.6 and BC = 9.2.
   Then AC = _____.

8. In the figure at right, suppose that AC = 15 and BC = 9.
   Then AB = _____.

Measuring Angles

9. The instrument used to measure an angle is called a(n) _____

10. On the protractor shown, find these measures:
    a. m ∠ DBC _____.
    b. m ∠ ABD _____.

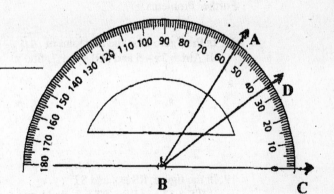

11. In the figure shown, find:
    a. m ∠ ABD if m ∠ ABC = 32° and m ∠ CBD = 29°. _____.
    b. m ∠ ABC if m ∠ ABD = 62° and m ∠ CBD = 28°. _____.
    c. m ∠ ABC if m ∠ ABD = 65° and $\overrightarrow{BC}$ bisects ∠ ABD. _____.

12. The sides of a(n) _____ angle form opposite rays.

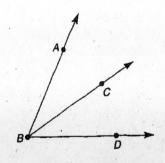

3

13. The sides of a(n) _____ angle form perpendicular rays.

Constructions

14. The instruments used to perform a construction are the straightedge and the _____.

15. The compass can be used to draw a circle or the part of a circle known as a(n) _____.

16. Construct a line segment that is congruent to $\overline{AB}$.

A   B

17. Use construction to locate the midpoint of $\overline{CD}$.

C   D

Further Problems

18. In the figure M is the midpoint of $\overline{AB}$.
     If AM = 3x − 5 and MB = x + 7, find x. _____

A   M   B

19. In the figure, RS = x and ST = y.
     If RT = 17 and RS − ST = 3, find x and y.

_____ .

R   S   T

4

Mathematical Systems

1. The four parts of a mathematical system are undefined terms, _____ (defined terms), postulates, and _____.

2. Some examples of mathematical systems are algebra, _____, and calculus.

A Good Definition

3. A good definition places the term being defined into a category. In the following definition, what is the category? "An isosceles triangle is a triangle with 2 congruent sides." _____.

4. A definition must be reversible. In the definition of congruent angles, state the reversible counterpart of the statement, "If 2 angles are congruent, then these angles have equal measures."
_____

Initial Postulates

5. Complete: Through two distinct points, there is exactly one _____.

6. Complete: (Ruler Postulate) The measure of any line segment is a unique _____ number.

7. Complete: (Segment-Addition Postulate) If X is on $\overline{AB}$ with A-X-B, then _____.

8. Complete: If two lines intersect, they intersect in a _____.

Definitions

9. If $\overline{AB}$ and $\overline{CD}$ are congruent, then _____.

A _____ B

C _____ D

10. Parallel lines are lines that lie in the same plane but do not _____.

Symbols

11. Draw a line segment to match the description with the symbol:
   a. line AB $\overrightarrow{AB}$
   b. line segment AB $\overleftrightarrow{AB}$
   c. length of line segment AB   AB
   d. ray AB $\overline{AB}$

12. Write each statement by making use of symbols:
   a. Line segment RS is congruent to line segment XY. _____.
   b. Lines r and s are parallel. _____.
   c. Ray AB is perpendicular to ray AC. _____.

Planes

13. Complete: Coplanar points lie in the same _____.

14. Complete: Horizontal planes R and S (not shown) must be _____.

15. Complete: Through 3 noncollinear points, there is exactly one _____.

16. Complete: If two planes intersect, their intersection is a(n) _____.

Further Problems

A ·

17. Consider the noncollinear points A, B, C, and D shown.   D ·
    By choosing two points (like A and B) at a time, what is   · B
    the total number of lines determined by these points? _____.
                                                                C ·

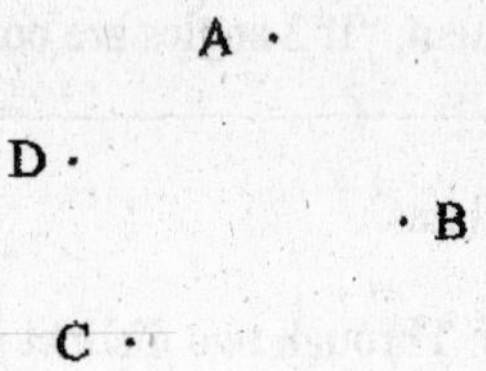

18. Points R, S, and T (not shown) are collinear with RS = 12 and ST = 17.
    What are the two possible lengths of $\overline{RT}$? _____ and _____.

Constructions

19. Given $\overline{XY}$, construct $\overline{XM}$ so that XM = $\frac{1}{2}$ (XY).

X ———————————— Y

20. Using $\overline{XY}$ as shown in Exercise 19, construct $\overline{XR}$
    so that XR = 2(XY).

6

Angle Postulates

1. Complete: (Protractor Postulate) The measure of an angle is a unique _____ number.

2. Complete: (Angle-Addition Postulate) If point D lies in the interior of ∠ ABC, then m ∠ ABD + m ∠ DBC = _____ .

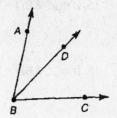

Types of Angles

3. An angle whose positive measure is less than 90^0 is a(n) _____ angle.

4. An angle whose measure is exactly 90^0 is a(n) _____ angle.

5. An angle whose measure is between 90^0 and 180^0 is a(n) _____ angle.

6. An angle whose positive measure is exactly 180^0 is a(n) _____ angle.

Pairs of Angles

7. Two angles are _____ if their measures are equal.

8. As shown, ∠ MNP and ∠ PNQ are _____ angles.

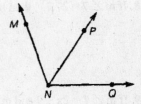

9. If ∠ MNP ≅ ∠ PNQ in the figure, then $\overrightarrow{NP}$ is said to _____ ∠ MNQ.

10. Two angles whose sum of measures is 90^0 are _____ angles.

11. Two angles whose sum of measures is 180^0 are _____ angles.

12. In the figure, ∠ 7 and ∠ 8 are _____ angles.

Constructions

13. Construct an angle congruent to ∠ RST.

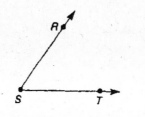

14. Construct the angle-bisector of ∠ RST.

7

Further Problems

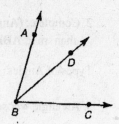

15. If m $\angle$ ABD = 29^0 and m $\angle$ DBC = 42^0, then m $\angle$ ABC = _____ .

16. If m $\angle$ ABD = 30.3^0 and m $\angle$ ABC = 72.6^0, then m $\angle$ DBC = _____ .

17. If m $\angle$ ABD = x, m $\angle$ DBC = 2x – 3, and m $\angle$ ABC = 72,
    then x = _____ .

18. If m $\angle$ 7 = 27.3^0, then m $\angle$ 8 = _____ and m $\angle$ 5 = _____ .

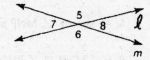

19. If m $\angle$ 7 = 2x + 9 and m $\angle$ 8 = 5x – 21, then x = _____ .

20. If m $\angle$ 7 = 2x - 11 and m $\angle$ 5 = 5(x + 6), then x = _____ .

Properties of Equality

1. Complete each property:
   a. Addition Property of Equality: If a = b, then a + c = _____ .
   b. Subtraction Property of Equality: If a = b, then _____
   c. Multiplication Property of Equality: If a = b, then _____ .

   d. Division Property of Equality: If a = b and c ≠ 0, then $\dfrac{a}{c}$ = _____ .

2. Use the property named to complete each conclusion:
   a. Addition Property of Equality: If 2x – 3 = 7, then _____ .
   b. Division Property of Equality: If 2x = 10, then _____ .

Further Algebraic Properties

3. Given that a, b, and c represent real numbers, name the Property of Equality illustrated:
   a. a(b + c) = ab + ac _____ .
   b. If a = b, then b can replace a in any equation. _____ .
   c. If a = b and b = c, then a = c. _____ .

4. Complete each statement:
   a. 2(x + 5) = _____ .
   b. With $\dfrac{1}{2}$ = 0.5 and 0.5 = 50%, it follows that _____ .

Algebraic Proof

5. In a proof, we make claims called _____ and justify these claims with reasons
   such as Given, definitions, postulates, and theorems.

6. Complete the *reasons* for the following algebraic proof.

   Given: 2(x – 5) + 3 = 17
   Prove: x = 12

   Proof

Statements	Reasons
(1.) 2(x – 5) + 3 = 17	(1.) _____
(2.) 2x – 10 + 3 = 17	(2.) _____
(3.) 2x – 7 = 17	(3.) _____
(4.) 2x = 24	(4.) _____
(5.) x = 12	(5.) _____

7. Complete the *statements* and *reasons* for the following algebraic proof.

   Given: 2x – 9 = 3x - 15
   Prove: x = 6 (same as 6 = x)

   Proof

Statements	Reasons
(1.) 2x – 9 = 3x - 15	(1.) _____
(2.) 2x + 6 = 3x	(2.) _____
(3.) _____	(3.) _____

9

8. In a geometric proof, the statements are written in a logical sequence and with each statement supported by an appropriate reason. Given the figure shown, name the reason that allows you to conclude that:

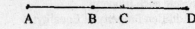

   a. AB + BC = AC _____. A   B   C   D
   b. If AB = CD, then AB + BC = BC + CD _____.

9. In the figure for Exercise 8, suppose that AB = 9, BC = 2, and CD = 9. Does AC = BD? _____.

10. Complete the *reasons* for this geometric proof. Refer to the drawing provided.

   Given: AB = CD on $\overline{AD}$
   Prove: AC = BD

   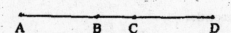

   A   B   C   D

   Proof

Statements	Reasons
(1.) AB = CD on $\overline{AD}$	(1.) _____
(2.) AB + BC = BC + CD	(2.) _____
(3.) AB + BC = AC and BC + CD = BD	(3.) _____
(4.) AC = BD	(4.) _____

11. Consider the drawing shown in the following exercise. Suppose that m∠RSW = 43°, m∠VSW = 21°, and m∠VST = 43°. Does it follow that m∠RSV = m∠WST? _____.

12. Complete the missing *statements* and *reasons* for the following geometric proof.

   Given: m∠RSW = m∠VST
   Prove: m∠RSV = m∠WST

   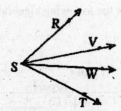

   Proof

Statements	Reasons
(1.) _____	(1.) _____
(2.) m∠RSW = m∠RSV + m∠VSW and m∠VST = m∠WST + m∠VSW	(2.) _____
(3.) m∠RSV + m∠VSW = m∠WST + m∠VSW	(3.) Substitution Property of Equality
(4.) _____	(4.) _____

10

Perpendicular Lines

1. If line m is vertical and line n is horizontal, how are coplanar lines m and n related? _____.

2. If 2 lines are perpendicular, these lines meet to form _____ angles.

Relations

3. The relationship "is perpendicular to" relates lines while the relation "is equal to" relates _____.

4. By name, the three special properties used to characterize a relation are:
   a. _____ Property: a R a
   b. _____ Property: If a R b, then b R a.
   c. _____ Property: If a R b and b R c, then a R c.

5. The Transitive Property for a relation, assuming one exists, can be used to relate the first object to the last object. Given that $\angle 1 \cong \angle 2$, $\angle 2 \cong \angle 3$, and $\angle 3 \cong \angle 4$, you may conclude _____.

6. When the relation "is congruent to" is used to relate angles, which properties (Reflexive, Symmetric, and Transitive) exist? _____.

7. When the relation "is greater than" is used to relate numbers, which properties (Reflexive, Symmetric, and Transitive) exist? _____.

8. When the relation "is perpendicular to" is used to relate lines, which properties (Reflexive, Symmetric, and Transitive) exist? _____.

9. When the relation "is equal to" is used to relate numbers, which properties (Reflexive, Symmetric, and Transitive) exist? _____.

Constructions

10. a. Given point P on line t in plane Q, how many lines can be drawn in plane Q that are perpendicular to line t at point P? _____.

    b. Given point P on line t, how many lines can be drawn in space that are perpendicular to line t at point P? _____.

11. Construct the line that is perpendicular to $\overleftrightarrow{AB}$ at point P.

$\xleftarrow{\hspace{2cm}} \bullet_A \quad\quad\quad \bullet_P \quad\quad\quad \bullet_B \xrightarrow{\hspace{1cm}}$

11

12. Construct the perpendicular-bisector of $\overline{AB}$.

A _____ B

Further Problems

13. Supply or complete missing *reasons* for the proof of the theorem,
    "If two lines are perpendicular, they meet to form a right angle."

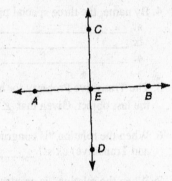

Given: $\overleftrightarrow{AB} \perp \overleftrightarrow{CD}$, intersecting at point E
Prove: $\angle AEC$ is a right angle

Proof

Statements	Reasons
(1.) $\overleftrightarrow{AB} \perp \overleftrightarrow{CD}$, intersecting at point E	(1.) _____
(2.) $\angle AEC \cong \angle CEB$	(2.) _____ lines form $\cong$ adjacent angles.
(3.) m $\angle AEC$ = m $\angle CEB$	(3.) _____
(4.) $\angle AEB$ is a st. angle, so m $\angle AEB = 180^0$	(4.) _____
(5.) m $\angle AEC$ + m $\angle CEB$ = m $\angle AEB$	(5.) _____
(6.) m $\angle AEC$ + m $\angle CEB = 180^0$	(6.) _____
(7.) m $\angle AEC$ + m $\angle AEC = 180^0$, so $2 \cdot$ m $\angle AEC = 180^0$	(7.) _____
(8.) m $\angle AEC = 90^0$	(8.) _____
(9.) $\angle AEC$ is a right angle	(9.) _____

14. In Example 3 of Section 1.6, we verify the theorem:
    "If two lines intersect, the vertical angles formed are congruent."
    Use information from this theorem to solve each problem.

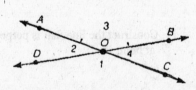

    a. If m $\angle 1 = 142^0$, then m $\angle 3$ = _____.
    b. If m $\angle 1 = x$, then m $\angle 3$ = _____.
    c. If m $\angle 1 = x$, then m $\angle 2$ = _____.
    d. If m $\angle 2 = 2x + 3$ and m $\angle 4 = 5x - 33$, then x = _____.

12

Hypothesis and Conclusion

1. In the conditional statement "If P, then Q," the hypothesis is the simple statement ____ and the conclusion is the simple statement _____.

2. In the following statement, underline the hypothesis once and the conclusion twice.
   "If two lines intersect, then the vertical angles formed are congruent."

3. Rewrite the following statement in the form of the conditional statement "If P, then Q."
   "All isosceles triangles have a pair of congruent sides."
   _____

The Written Parts of a Formal Proof

4. The five parts that must be shown in the formal proof of a theorem are:
   a. Statement of the _____.
   b. Drawing representing the facts found in the _____ of the theorem.
   c. Given, which describes the drawing based upon the _____ of the theorem.
   d. Prove, which describes the drawing based upon the _____ of the theorem.
   e. Proof, which provides statements and reasons in a logical order – beginning with the _____ statement and ending with the _____ statement.

5. For the stated theorem and the related drawing, write the Given and Prove.
   "If two lines meet to form a right angle, then these lines are perpendicular."
   Given: _____.
   Prove: _____.

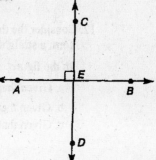

Converse of a Theorem

6. The converse of the statement "If P, then Q" is the statement _____.

7. Is the converse of every conditional statement "If P, then Q" necessarily true? _____.

8. a. Write the converse of the statement, "If two lines are perpendicular, they meet to form a right angle." _____
   b. Is the converse that was written in part (a) true or false? _____.

Theorems of Section 1.7

9. Suppose that (a) ∠ 1 is complementary to ∠ 3
   and that (b) ∠ 2 is also complementary to ∠ 3.
   How are ∠ 1 and ∠ 2 related? _____.

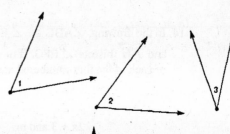

10. Consider the theorem, "If the exterior sides of two adjacent acute
    angles form perpendicular rays, then these angles are complementary."
    In the drawing, $\overrightarrow{BA} \perp \overrightarrow{BC}$. Given that m ∠ 1 = 28°, use the
    theorem to find m ∠ 2. _____.

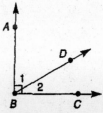

13

11. Supply or complete missing *statements* or *reasons* for the proof of the theorem,
    "If two angles are supplementary to the same angle, then these angles are congruent."

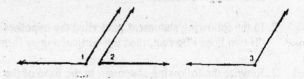

Given: $\angle 1$ is supplementary to $\angle 2$;
       $\angle 3$ is supplementary to $\angle 2$
Prove: $\angle 1 \cong \angle 3$

Proof

Statements	Reasons
(1.) $\angle 1$ is supplementary to $\angle 2$; $\angle 3$ is supplementary to $\angle 2$	(1.) _____.
(2.) $m\angle 1 + m\angle 2 = 180$; $m\angle 3 + m\angle 2 = 180$	(2.) _____
(3.) $m\angle 1 + m\angle 2 =$ _____	(3.) Substitution Property of Equality
(4.) $m\angle 1 = m\angle 3$	(4.) _____ Property of Equality
(5.) _____	(5.) _____

12. Consider the theorem, "If the exterior sides of two adjacent angles
    form a straight line, then these angles are supplementary."

    In the figure, $\overleftrightarrow{EG}$ is a straight line.

    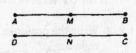

    a. Given that $m\angle 3 = 128^0$, find $m\angle 4$. _____
    b. Given that $m\angle 4 = 49^0$, find $m\angle 3$. _____
    c. Given that $m\angle 3 = 3y$ and $m\angle 4 = y$, find y. _____

13. In the figure, $\overline{AB} \cong \overline{DC}$. M is the midpoint of $\overline{AB}$
    and N is the midpoint of $\overline{DC}$. How are the four line segments
    $\overline{AM}$, $\overline{MB}$, $\overline{DN}$, and $\overline{NC}$ related? _____.

    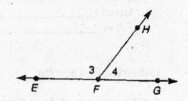

14. In the drawing, $\angle ABC \cong \angle EFG$. Also, $\overrightarrow{BD}$ bisects $\angle ABC$
    and $\overrightarrow{FH}$ bisects $\angle EFG$. If $m\angle ABC = 56^0$, find the measure
    of each of the four numbered angles. _____.

    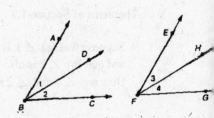

14

Parallel and Perpendicular Lines

1. In the drawing, how many lines can be drawn through
   point P that are perpendicular to line ℓ? _____.

2. In the drawing, how many lines can be drawn through
   point P that are parallel to line ℓ? _____.

3. In a plane, line m is drawn so that it is perpendicular to line t. In the
   same plane, a second line n is drawn so that it is also perpendicular to t.
   How are lines m and n related? _____.

Lines with a Transversal

4. In the figure, use numbers to name four angles
   that are interior angles. _____.

5. In the figure, use numbers to name two angles
   that are corresponding angles. _____.

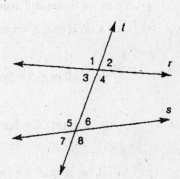

6. In the figure, use numbers to name two angles
   that are alternate interior angles. _____.

Parallel Lines and Angles

7. Complete this postulate: If two _____ lines are cut by a transversal,
   then the corresponding angles are congruent.

8. Assume that $\ell \parallel$ m in the figure. Which angle is the "alternate
   interior angle" that is congruent to $\angle$ 3? _____.

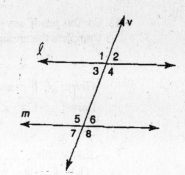

9. Complete this theorem: If two parallel lines are cut by a transversal, then the
   exterior angles on the same side of the transversal are _____.

10. Complete the missing statements and reasons for the proof of the theorem:
    "If two parallel lines are cut by a transversal, then the alternate interior angles are congruent."

Given: a $\parallel$ b; transversal k

Prove: $\angle 3 \cong \angle 6$

Proof

Statements	Reasons
(1.) _____	(1.) _____
(2.) $\angle 2 \cong \angle 6$	(2.) If two $\parallel$ lines are cut by a transversal, then _____ angles are $\cong$.
(3.) $\angle 3 \cong \angle 2$	(3.) If two lines intersect, the _____ angles formed are congruent.
(4.) _____	(4.) Transitive Property of Congruence

15

Further Problems

11. In the figure, $\ell \parallel m$ and transversal v. Find:

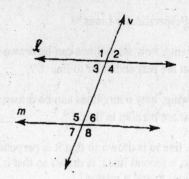

    a. $m\angle 5$ if $m\angle 1 = 118^0$. _____.

    b. $m\angle 5$ if $m\angle 8 = 115^0$. _____.

    c. $m\angle 2$ if $m\angle 8 = 117.5^0$. _____.

    d. $m\angle 2$ if $m\angle 1 = 3(m\angle 2)$ _____

12. In the figure, $a \parallel b$ and transversal k. Find:

    a. x, if $m\angle 5 = x$ and $m\angle 1 = 42^0$. _____.

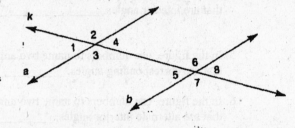

    b. y, if $m\angle 5 = 3y + 13$ and $m\angle 8 = 4y + 3$. _____.

    c. z, if $m\angle 2 = 5(z + 3)$ and $m\angle 8 = 7(z - 5)$ _____.

    d. w, if $m\angle 3 = 7w - 10$ and $m\angle 5 = 70 - w$. _____.

13. For this proof, any postulate or theorem of Section 2.1 can be cited as a reason.
Complete the missing *statements* and *reasons* for the following proof problem.

Given: $\ell \parallel m$ and $m \parallel n$
Prove: $\angle 1 \cong \angle 4$

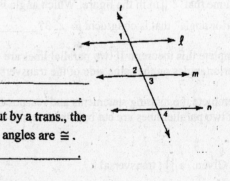

Proof

Statements	Reasons
(1.) $\ell \parallel m$	(1.) _____.
(2.) $\angle 1 \cong \angle 2$	(2.) If 2 $\parallel$ lines are cut by a trans., the _____ angles are $\cong$.
(3.) $\angle 2 \cong \angle 3$	(3.) _____.
(4.) _____	(4.) Given
(5.) $\angle 3 \cong \angle 4$	(5.) _____.
(6.) _____	(6.) Trans. Prop. for Congruence of Angles

Statements Written in Symbolic Form

1. Given that the symbolic form of the statement "If P, then Q" is written P → Q while "not P" is
   written ~ P, write each statement in symbolic form:
   a. Converse of "If P, then Q" _____.
   b. Inverse of "If P, then Q" _____.
   c. Contrapositive of "If P, then Q" _____.

2. Given that the conditional statement P → Q is true, which statement (converse, inverse, or
   contrapositive) *must* also be true? _____.

Law of Negative Inference (Law of Contraposition)

3. Write the conclusion of the Law of Negative Inference.
   (1) P → Q
   (2) ~ Q
   (C) _____.

4. Use the Law of Negative Inference to state the conclusion of the following argument.
   (1) If Pablo lives in Guadalajara, then he lives in Mexico.
   (2) Pablo does not live in Mexico.
   (C) _____.

When to use Indirect Proof

5. Indirect proof is often used when statement _____ of the theorem "If P, then Q" has the form
   of a negation.

6. For which of the following "Prove" statements would you be likely to use an indirect proof?
   a. c ≠ d   b. c = d   c. ∠1 and ∠2 are not congruent   _____.

7. Indirect proof is also used to prove uniqueness theorems. For which of the following theorems
   would you be likely to use an indirect proof? _____.
   a. The midpoint of a line segment is unique.
   b. An angle has exactly one bisector.
   c. Through a point not on a line, there is only one line parallel to the given line.
   d. All right angles are congruent.

The Method of Indirect Proof

8. To use indirect proof with the proof problem of the form
   Given: P
   Prove: Q,
   we use these steps:
   a. Suppose that statement _____ is true.
   b. Reason from the supposed statement in (a) until you reach a _____.
   c. With a contradiction in (b), note that the supposition in (a) must be false and it follows that
      statement _____ must therefore be true.

17

9. Fill the blanks to complete the proof of the following theorem:
   "If ∠ 1 is not congruent to ∠ 2, then ∠ 1 and ∠ 2 are not vertical angles."
   Note: No figure is provided!

   Given: ∠ 1 is not congruent to ∠ 2
   Prove: ∠ 1 and ∠ 2 are not vertical angles

   Proof: Suppose that ∠ 1 and ∠ 2 are _____.
   Then _____ because vertical angles are congruent.
   However, it is Given that ∠ 1 is not congruent to ∠ 2; thus, we have a
   contradiction of a known fact. It follows that the supposition must be _____.
   so it follows that _____.

10. Fill the blanks to complete the proof of the following theorem:

    "The bisector of a given angle is unique."

        Given: $\overrightarrow{BD}$ bisects ∠ ABC
        Prove: $\overrightarrow{BD}$ is the only angle-bisector for ∠ ABC

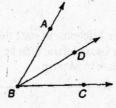 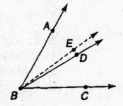

    Proof: $\overrightarrow{BD}$ bisects ∠ ABC, so m ∠ ABD = $\frac{1}{2}$ (_____)

    Suppose that $\overrightarrow{BE}$ also bisects ∠ ABC, so m ∠ ABE = $\frac{1}{2}$ (m ∠ ABC).

    By the Angle-Addition Postulate, m ∠ ABD = m ∠ ABE + m _____.

    Using the Substitution property of Equality, the equation in the preceding

    statement can be written $\frac{1}{2}$ m ∠ ABC = $\frac{1}{2}$ m ∠ ABC + _____.

    Applying the Subtraction Property of Equality, m ∠ EBD = _____ ; but this
    equation contradicts the Protractor Postulate, which states that the measure
    of any angle is a unique _____ number. Based upon the contradiction
    of the stated postulate, the supposition must be false and it follows that

    _____.

A Contrast

1. In Section 2.1, the postulates and theorems have the general form "If two parallel lines are cut by a transversal, then . . ." Those postulates and theorems can be used to justify that a pair of angles are _____ or else that a pair of angles are _____.

2. In this section (Section 2.3), the theorems take the form, "If . . . , then the lines are parallel." These statements can be used to justify that _____.

Lines with a Transversal

3. Complete: If two lines are cut by a transversal so that corresponding angles are congruent, then these lines are _____.

4. Complete: If two lines are cut by a transversal so that the interior angles on the same side of the transversal are _____, then these lines are parallel.

5. Complete: If two lines are each parallel to a third line, then these lines are _____.

6. Complete: If two _____ lines are each perpendicular to a third line, then these lines are parallel.

Applications

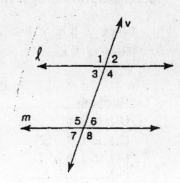

7. If m∠1 = 107°, find m∠5 so that ℓ ‖ m. _____.

8. If m∠4 = 106°, find m∠6 so that ℓ ‖ m. _____.

9. If m∠2 = 72 and m∠7 = 4x + 20, find x so that ℓ ‖ m. _____.

10. If m∠3 = 2x + 26 and m∠5 = 6(x − 1), find x so that ℓ ‖ m. _____.

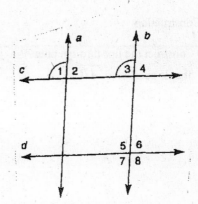

11. In the figure, which lines (if any) must be parallel if ∠1 ≅ ∠3? _____.

12. In the figure, which lines (if any) must be parallel if ∠1 ≅ ∠4? _____.

13. In the figure, which lines (if any) must be parallel if ∠8 ≅ ∠3? _____.

14. In the figure, which lines (if any) must be parallel if ∠3 and ∠7 are supplementary? _____.

Proving Lines Parallel

15. Complete the missing *statements* and *reasons* for the proof of the theorem:
"If two lines are cut by a transversal so that alternate interior angles are
congruent, then these lines are parallel."

Given: Lines ℓ and m; transversal t;
$\qquad \angle 2 \cong \angle 3$

Prove: $\ell \parallel m$

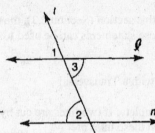

Proof

Statements	Reasons
(1.) Lines ℓ and m; trans t; $\angle 2 \cong \angle 3$	(1.) _____
(2.) $\angle 1 \cong \angle 3$	(2.) _____
(3.) _____	(3.) Transitive Prop. of Congruence
(4.) _____	(4.) If 2 lines are cut by a trans. so that corr. angles are $\cong$, these lines are parallel.

16. Complete the missing *statements* and *reasons* for this proof.

Given: $\ell \parallel m$; $\angle 3 \cong \angle 4$

Prove: $\ell \parallel n$

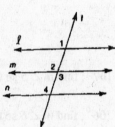

Proof

Statements	Reasons
(1.) $\ell \parallel m$	(1.) _____
(2.) $\angle 1 \cong \angle 2$	(2.) _____
(3.) $\angle 2 \cong \angle 3$	(3.) _____
(4.) _____	(4.) Given
(5.) $\angle 1 \cong \angle 4$	(5.) Transitive Prop. of Congruence (for angles)
(6.) _____	(6.) If 2 lines are cut by a trans. so that corr. angles are $\cong$, these lines are parallel.

A Construction

17. Construct the line through point P
that is parallel to $\overleftrightarrow{AB}$.

$\bullet P$

A   B

Triangle Types Classified by Sides

1. If no two sides of a triangle are congruent, the triangle is a(n) _____ triangle.

2. If exactly two sides of a triangle are congruent, the triangle is a(n) _____ triangle.

3. If all three sides of a triangle are congruent, the triangle is a(n) _____ triangle.

Triangle Types Classified by Angles

4. If all three angles of a triangle are acute, the triangle is a(n) _____ triangle.

5. If one angle of a triangle is a right angle, the triangle is a(n) _____ triangle.

6. If one angle of a triangle is obtuse, the triangle is a(n) _____ triangle.

7. If all three angles of a triangle are congruent, the triangle is a(n) _____ triangle.

The Sum of Interior Angles of a Triangle

8. In the figure, $m\angle 1 + m\angle 2 + m\angle 3 =$ _____ degrees.

9. In the figure, $\overleftrightarrow{ED} \parallel \overline{AB}$. Then $\angle 1 \cong \angle A$ and
   $\angle 3 \cong \angle B$ because these pairs of angles are
   _____ _____ angles.

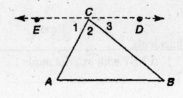

10. The sum of the three interior angles of Δ ABC
    is _____ degrees.

Corollaries of the theorem: "The sum of measures of the interior angles of a triangle is 180^{0}."

11. Complete: Each interior angle of an equiangular triangle measures _____ degrees.

12. Complete: The acute angles of a right triangle are _____.

13. Complete: If two angles of one triangle are congruent to two angles of a second triangle,
    then the _____ angles are also congruent.

14. Complete: The measure of a(n) _____ angle of a triangle equals the sum of the
    measures of the two nonadjacent interior angles.

Problems

15. In the figure, $m\angle 1 + m\angle 2 =$ _____.

16. In the figure, find x if $m\angle 1 = 4x + 7$
    and $m\angle 2 = 2x + 3$. _____

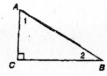

21

Further Problems

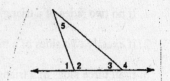

17. In the figure, find m $\angle$ 4 if m $\angle$ 2 = 113^0
    and m $\angle$ 5 = 26^0. _____.

18. Find an expression for m $\angle$ 1 if m $\angle$ 5 = x
    and m $\angle$ 4 = y. _____.

A Sample Proof

19. The following proof is that of the corollary,

    "The measure of an exterior angle of a triangle equals
    the sum of measures of the two nonadjacent interior angles."

    Complete the missing *statements* and *reasons* for the following proof.

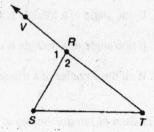

Given: Δ RST with exterior angle 1
Prove: m $\angle$ 1 = m $\angle$ S + m $\angle$ T

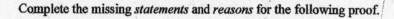

Proof

Statements	Reasons
(1.) Δ RST with exterior angle 1	(1.) _____.
(2.) _____	(2.) If the exterior sides of 2 adjacent angles form a straight line, these angles are supplementary.
(3.) m $\angle$ 1 + m $\angle$ 2 = 180^0	(3.) _____
(4.) m $\angle$ 2 + m $\angle$ S + m $\angle$ T = 180^0	(4.) _____.
(5.) _____	(5.) Substitution Property of Equality
(6.) _____	(6.) Subtraction Property of Equality

Polygons

1. Give the name of the polygon that has:
   a. 3 sides _____
   b. 4 sides _____.
   c. 5 sides _____
   d. 6 sides _____.
   e. 8 sides _____
   f. 10 sides _____.

2. What word is used to describe a polygon that has:
   a. all sides congruent? _____.
   b. all angles congruent? _____.
   c. all sides congruent and all angles congruent? _____.

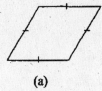

(a)   (b)   (c)

Diagonals of a Polygon

3. Name the two diagonals of quadrilateral ABCD (not shown). _____.

4. By drawing and counting, how many diagonals does a pentagon have? _____.

5. Where n is the number of sides for a given polygon, the polygon has $D = \dfrac{n(n-3)}{2}$ diagonals.

   Find the total number of diagonals in an octagon. _____.

Sum of the Interior Angles of a Polygon

6. For a polygon with n sides, diagonals can be drawn from a selected vertex to separate the polygon into _____ triangles.

7. The sum of measures of the interior angles of a polygon with n sides is S = _____.

8. Find the sum of the measures of the interior angles for any quadrilateral. _____.

9. Find the sum of the measures of the interior angles for a polygon with 12 sides. _____.

Regular and Equiangular Polygons

10. Because the sum of measures of the interior angles of a polygon with n sides is $(n-2) \cdot 180^0$, the formula for the measure of each interior angle of a regular (or equiangular) polygon is I = _____.

11. Find the measure of each interior angle of a(n):
    a. equiangular triangle. _____.
    b. regular pentagon. _____.
    c. equiangular polygon with 10 sides (a decagon). _____.

12. For a regular octagon, find the measure of:
    a. each interior angle. _____.
    b. each exterior angle. _____.

Exterior Angles of a Polygon

13. The sum of measures of the exterior angles (one at each vertex) of a polygon of n sides is
    always _____ degrees.

14. Find the measure of each exterior angle of a(n):
    a. regular decagon. _____.
    b. equiangular polygon with 18 sides. _____.

Polygrams

15. Informally, a _____ is a star-shaped figure with 5 or more vertices.

16. Being specific as to type, name each regular polygram shown below:

    a. _____.

    b. _____.

    c. _____.

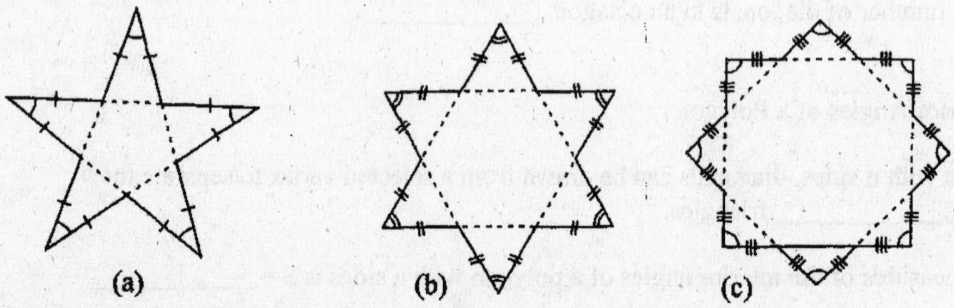

    (a)   (b)   (c)

17. Consider the regular pentagram shown in 16(a).
    a. How many congruent acute (interior) angles are there? _____.
    b. How many congruent reflex (interior) angles are there? _____.
    c. How many congruent sides are there? _____.

24

Line Symmetry

1. When a figure has horizontal symmetry, there is a vertical axis (line) of symmetry, Similarly, a figure that has vertical symmetry has a _____ axis (line) of symmetry.

2. Of the geometric figures shown below, which one(s) have line symmetry? _____.
   (a) Square   (b) Letter A   (c) Parallelogram   (d) Letter S

   A   S

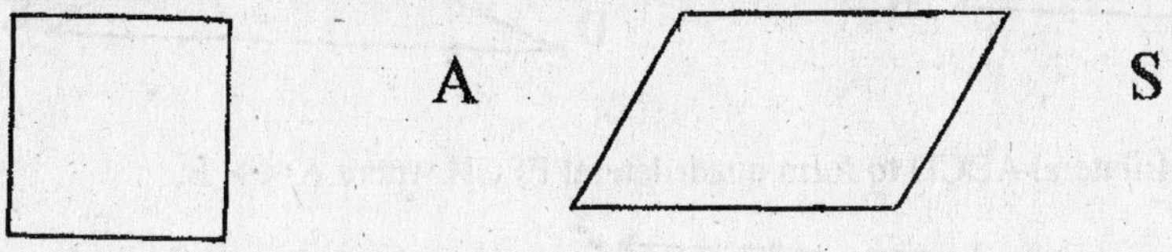

3. Of the figures shown in Exercise 2, name any that have more than one line of symmetry? _____.

4. Each figure shown has line symmetry. Is the line of symmetry vertical, horizontal, both, or neither?
   (a) _____ ; (b) _____ (c): _____ (d): _____

   (a) Isosceles Trapezoid   (b) Letter E   (c) Rectangle   (d) Letter T

   E   T

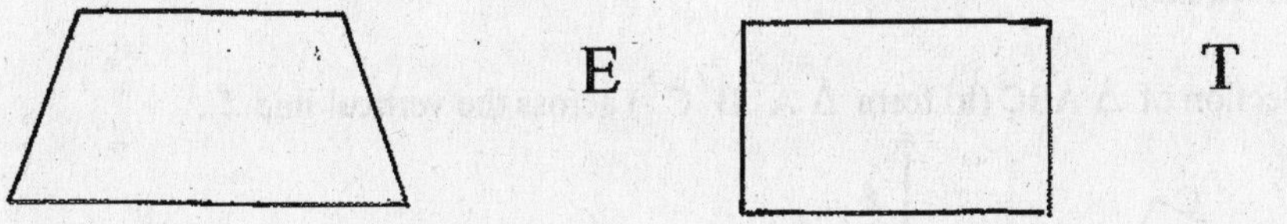

Point Symmetry

5. When a figure has point symmetry with respect to point P, there is a point C on the figure that corresponds to a point A on the figure in such a way that P is the _____ of $\overline{AC}$.

6. Of the geometric figures shown below, which one(s) have point symmetry? _____
   (a) Square   (b) Letter A   (c) Parallelogram   (d) Letter S

   A   S

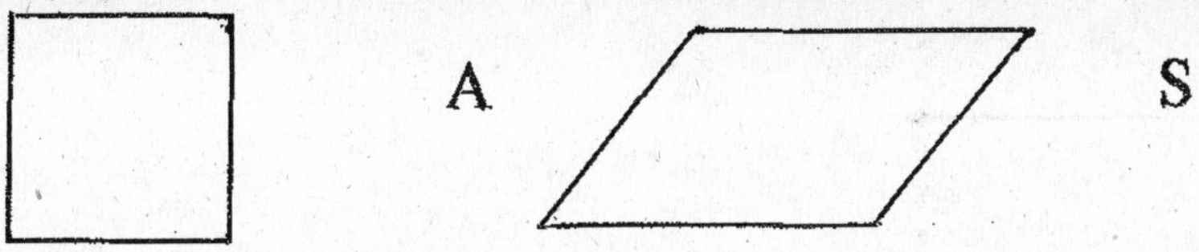

7. Do all regular polygons have point symmetry? _____

8. Of the figures shown, which one(s) have point symmetry? _____

   (a) Isosceles Trapezoid   (b) Letter E   (c) Rectangle   (d) Letter T

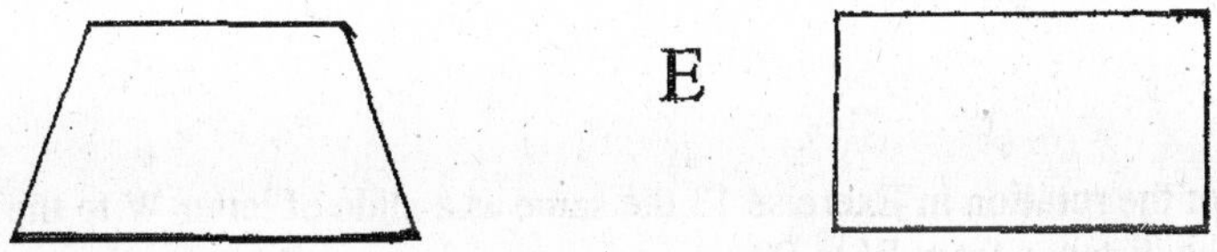

   E   T

25

Transformations (Slides)

9. When we slide △ ABC to form image △ DEF, we know that these triangles are _____.

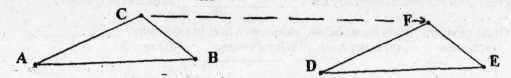

10. Complete the slide of quadrilateral ABCD to form quadrilateral EFGH where A ↔ E.

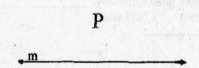

Transformations (Reflections)

11. Complete the reflection of △ ABC (to form △ A′B′C′) across the vertical line ℓ.

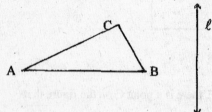

12. Complete the reflection of the letter P across the horizontal line m.

P

m

Transformations

13. Complete the clockwise rotation of the letter W
   through a 180⁰ angle about point P.   W   ·P

14. Is the effect (image produced) of the rotation in Exercise 13 the same as a slide of letter W to the right with a distance equal to twice the distance from W to P? _____.

26

Congruent Triangles

1. Given the statement △ ABC ≅ △ DEF, it follows that:
   a. ∠ A of △ ABC corresponds to _____ of △ DEF.
   b. Vertex C of △ ABC corresponds to _____ of △ DEF.
   c. Side _____ of △ ABC corresponds to $\overline{DF}$ of △ DEF.

2. Complete these properties for "congruence of triangles."
   a. Reflexive: △ ABC ≅ _____ .
   b. Symmetric: If △ ABC ≅ △ DEF, then _____ .
   c. Transitive: If △ ABC ≅ △ DEF and △ DEF ≅ △ GHK, then _____ .

Included Sides and Included Angles

3. In △ UTV, which:
   a. side is included by ∠ U and ∠ V? _____ .
   b. angle is included by sides $\overline{UV}$ and $\overline{UT}$? _____ .

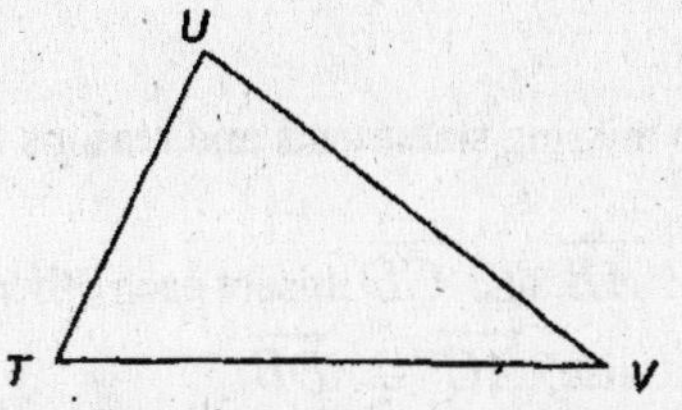

Reflexive property of Congruence

4. In a proof, the statement ∠ A ≅ _____ can be justified by the one-word reason Identity.

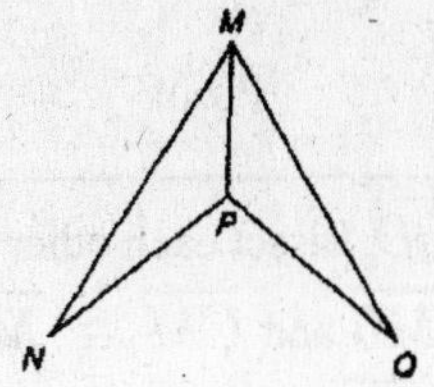

5. In the figure, you want to prove that △ NMP ≅ △ QMP.
   What statement can you write that is supported by the
   reason Identity? _____ .

6. In the figure, you wish to prove that △ ABC ≅ △ DBC.
   What statement can you write that is supported by the
   reason Identity? _____ .

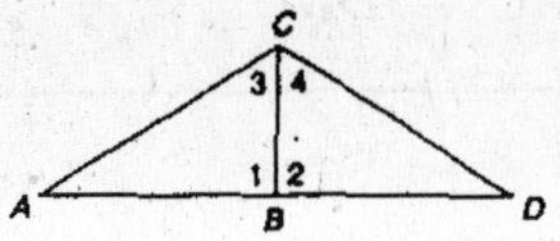

Methods Used to Prove that Two Triangles are Congruent

7. The postulate, "If the three sides of one triangle are congruent to the three sides of a second triangle, then the triangles are congruent" is indicated (in a proof) by the reason _____ .

8. When SAS is used to verify that triangles are congruent, the congruent angle(s) cited must be included by the pairs of _____ sides.

9. The four methods found in Section 3.1 that are used to prove that two triangles are congruent are SSS, SAS, _____, and _____ .

10. The figure shown illustrates why there is
    no method for proving two triangles congruent
    that is abbreviated _____ .

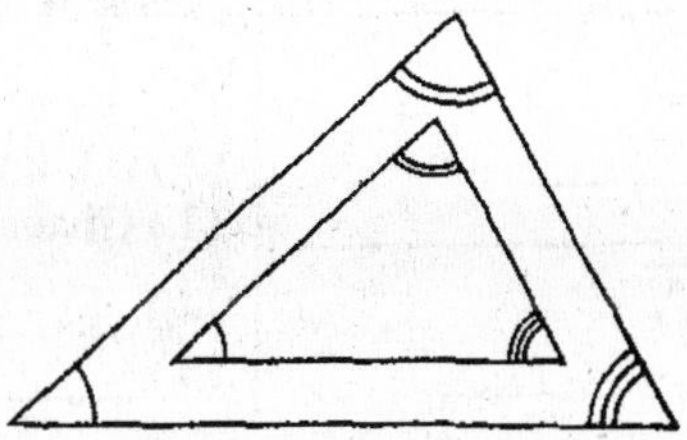

27

11. The figure shown illustrates why there is no method for proving triangles congruent that is abbreviated _____.

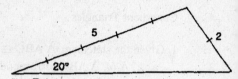

Proving Triangles Congruent

12. With congruent parts of triangles so marked, which method (SSS, SAS, ASA, or AAS) would be used to show that the two triangles are congruent?

(a) _____   (b) _____

   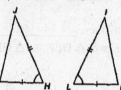

13. Provide missing statements and reasons for the following proof.

Given: $\overline{AB}$ and $\overline{CD}$ bisect each other at M;
  also, $\overline{AC} \cong \overline{DB}$
Prove: $\triangle AMC \cong \triangle BMD$

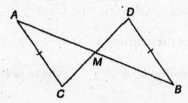

Proof

Statements	Reasons
(1.) $\overline{AB}$ and $\overline{CD}$ bisect each other at M	(1.) _____
(2.) $\overline{AM} \cong \overline{MB}$ and $\overline{CM} \cong \overline{MD}$	(2.) _____
(3.) $\overline{AC} \cong \overline{DB}$	(3.) _____
(4.) _____	(4.) _____

14. Provide missing statements and reasons for the following proof.

Given: $\overline{PN} \perp \overline{MQ}$; $\overline{MN} \cong \overline{NQ}$
Prove: $\triangle PNM \cong \triangle PNQ$

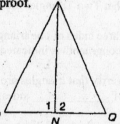

Proof

Statements	Reasons
(1.) $\overline{PN} \perp \overline{MQ}$	(1.) _____
(2.) $\angle 1 \cong$ _____	(2.) If 2 lines are perp., they form $\cong$ adj. angles.
(3.) _____	(3.) Given
(4.) $\overline{PN} \cong \overline{PN}$	(4.) _____
(5.) _____	(5.) _____

28

Corresponding Parts of Congruent Triangles

1. Based upon the definition of congruent triangles, "Corresponding parts of congruent triangles are congruent" is expressed by the abbreviation _____.

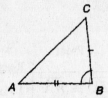
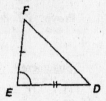

2. In the figure, $\triangle ABC \cong \triangle DEF$ by the reason SAS.
   Using CPCTC, we know that $\overline{AC} \cong$ _____.

3. In the figure, $\triangle ABC \cong \triangle DEF$ by the reason SAS.
   Using CPCTC, we know that $\angle A \cong$ _____.

Proof Using CPCTC

4. In order to use CPCTC as the reason for proving a pair of corresponding angles (or sides) of two triangles congruent, one must first show that the two triangles are _____.

5. Complete the missing statements and reasons for the following proof problem.

   Given: $\overline{WZ}$ bisects $\angle TWV$; $\overline{WT} \cong \overline{WV}$
   Prove: $\overline{TZ} \cong \overline{VZ}$

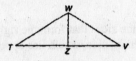

Proof

Statements	Reasons
(1.) $\overline{WZ}$ bisects $\angle TWV$	(1.) _____
(2.) _____	(2.) The bisector of an angle separates the angle into two congruent triangles.
(3.) $\overline{WT} \cong \overline{WV}$	(3.) _____
(4.) _____	(4.) Identity
(5.) $\triangle TWZ \cong \triangle VWZ$	(5.) _____
(6.) _____	(6.) CPCTC

6. If the "Prove" statement of the proof in Exercise 5 had read, "Prove: Z is the midpoint of $\overline{TV}$," we would have needed to add a seventh step. Complete this step.

Statements	Reasons
(7.) _____	(7.) _____

Suggestions for Proving Triangles Congruent

7. Some suggestions for proving triangles congruent include marking corresponding parts in a like manner. Mark the figures in a systematic manner, using:
   (a.) a _____ in the opening of each right angle;
   (b.) the same number of dashes on the _____ sides; and
   (c.) the same number of arcs in openings of congruent _____.

8. Consider the markings in the figure at the right.
   (a.) What type of angle is $\angle WZX$? _____.
   (b.) Which side of $\triangle ZXY$ is congruent to $\overline{WZ}$ in $\triangle XWZ$? _____.

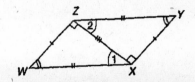

29

9. Complete the missing statements and reasons for the following proof problem.

Given: $\overline{ZW} \cong \overline{YX}$; $\overline{ZY} \cong \overline{WX}$

Prove: $\overline{ZY} \parallel \overline{WX}$

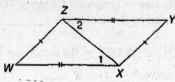

Proof

Statements	Reasons
(1.) $\overline{ZW} \cong \overline{YX}$; $\overline{ZY} \cong \overline{WX}$	(1.) _____
(2.) _____	(2.) Identity.
(3.) _____	(3.) SSS
(4.) $\angle 1 \cong \angle 2$	(4.) _____
(5.) _____	(5.) If two lines are cut by a transversal so that alt. int. angles are $\cong$, these lines are $\parallel$.

Right Triangles

10. In a right triangle, the sides that form the right angle are called the _____ of the right triangle. The remaining side, which lies opposite the right angle, is known as the _____ of the right triangle.

11. The method HL can only be used to prove that two _____ triangles are congruent.

12. The Pythagorean Theorem: In a right triangle, the square of length c of the hypotenuse is equal to the sum of squares of the lengths a and b of the legs of the right triangle; that is $c^2 =$ _____.

13. The Square Roots Property: Let x represent the length of a line segment and let p represent some positive number. If $x^2 = p$, then $x =$ _____.

14. In the right triangle shown, find:
   (a.) c, if a = 3 and b = 4 _____

   (b.) b, if c = 10 and a = 8 _____

   (c.) c, if a = 4 and b = 5 _____

   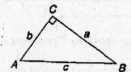

   (d.) a, if c = 12 and b = 7 _____

30

Segments, Rays, and Lines Related to the Triangle

1. If M is the midpoint of $\overline{BC}$, then $\overline{AM}$ is the
_____ of $\triangle$ ABC from vertex A to side $\overline{BC}$.

2. If M is the midpoint of $\overline{BC}$, then $\overleftrightarrow{FM}$ is the
perpendicular-bisector of side _____ of $\triangle$ ABC.

3. If $\angle$ BAD $\cong$ $\angle$ DAC, then $\overline{AD}$ is the angle-bisector
of _____ in $\triangle$ ABC.

4. With E on $\overline{BC}$ and $\overline{AE}$ $\perp$ $\overline{BC}$, $\overline{AE}$ is known as the
_____ of $\triangle$ ABC from vertex A to side $\overline{BC}$.

5. Every triangle has three angle-bisectors, _____ medians, _____ altitudes, and
three _____ of the sides of the triangle.

6. Complete: Corresponding altitudes of congruent triangles are _____ .

Isosceles Triangles

7. In an isosceles triangle, the two sides of equal length are the _____ of the isosceles triangle and the
remaining side is known as the _____ of the isosceles triangle.

8. In $\triangle$ MNP, $\overline{MP}$ $\cong$ $\overline{PN}$. For the isosceles triangle,
which side is the base of $\triangle$ MNP? _____ .

9. In $\triangle$ MNP, $\overline{MP}$ $\cong$ $\overline{PN}$. For the isosceles triangle,
which angle is the vertex angle of $\triangle$ MNP? _____ .

10. Provide missing statements and reasons for the proof of the theorem,
"The bisector of the vertex angle of an isosceles triangle separates the triangle into two congruent
triangles."

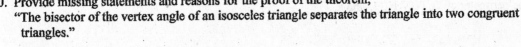

Given: $\triangle$ ABC with $\overline{AB}$ $\cong$ $\overline{BC}$; $\overline{BD}$ bisects $\angle$ ABC
Prove: $\triangle$ ABD $\cong$ $\triangle$ CBD

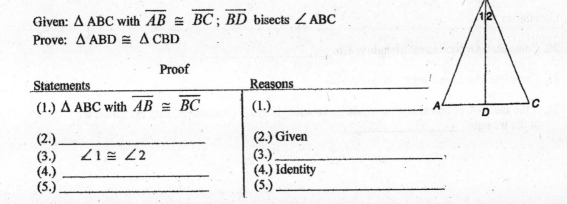

	Proof
Statements	Reasons
(1.) $\triangle$ ABC with $\overline{AB}$ $\cong$ $\overline{BC}$	(1.) _____
(2.) _____	(2.) Given
(3.) $\angle$ 1 $\cong$ $\angle$ 2	(3.) _____
(4.) _____	(4.) Identity
(5.) _____	(5.) _____

31

11. Complete this theorem: If two sides of a triangle are congruent,
    then the angles that lie _____ these sides are congruent.

12. Complete this theorem: If two angles of a triangle are congruent,
    then the sides that lie _____ these angles are congruent.

Select Problems

13. In △ TUV, $\overline{TV} \cong \overline{UV}$. Which angles of △ TUV
    are congruent? _____.

14. In △ TUV, ∠ T $\cong$ ∠ U. Which sides of △ TUV
    are congruent? _____.

15. In △ TUV, $\overline{TV} \cong \overline{UV}$. If m∠ T = 68°, find:
    a. m∠ U. _____   b. m∠ V. _____.

16. In △ TUV, $\overline{TV} \cong \overline{UV}$. If m∠ V = 54°, find:
    a. m∠ T. _____   b. m∠ U. _____.

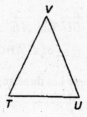

17. In △ TUV, $\overline{TV} \cong \overline{UV}$. If m∠ T = x + 12 and m∠ V = x, find: Exercises 13-19
    a. x. _____   b. m∠ T. _____.

18. In △ TUV, $\overline{TV} \cong \overline{UV}$. If TV = 12 and TU = 9, find
    the perimeter of △ TUV. _____.

19. In △ TUV, $\overline{TV} \cong \overline{UV}$. If TU = x and TV = 2x - 3, and
    the perimeter of △ TUV is 39, find x. _____.

Corollaries

20. Complete: An equilateral triangle is also _____.

21. Complete: An equiangular triangle is also _____.

22. If the measure of one side of an equiangular triangle is 4.7 centimeters, find the perimeter
    of the triangle. _____.

32

Basic Construction Methods Justified

1. Complete the missing statements and reasons for this proof, which justifies the
   method for construction of an angle congruent to a given angle.

Given: ∠ABC with
$$\overline{BD} \cong \overline{BE} \cong \overline{ST} \cong \overline{SR}$$
and $\overline{DE} \cong \overline{TR}$
Prove: ∠B ≅ ∠S

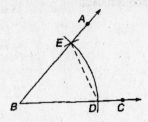
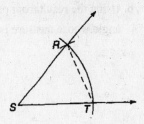

Proof

Statements	Reasons
(1.) ∠ABC with $\overline{BD} \cong \overline{BE} \cong \overline{ST} \cong \overline{SR}$	(1.) _____
(2.) _____	(2.) Given
(3.) △ EBD ≅ △ RST	(3.) _____
(4.) _____	(4.) _____

2. Construct an isosceles triangle ABC with vertex angle A
   as shown and with legs the length of $\overline{AB}$.

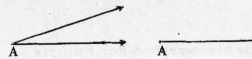

3. Complete the missing statements and reasons for this proof, which justifies the
   method for construction of the bisector of a given angle.

Given: ∠XYZ with $\overline{YM} \cong \overline{YN}$
and $\overline{MW} \cong \overline{NW}$
Prove: $\overrightarrow{YW}$ bisects ∠XYZ

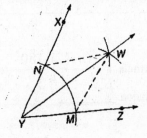

Proof

Statements	Reasons
(1.) ∠XYZ with $\overline{YM} \cong \overline{YN}$ and $\overline{MW} \cong \overline{NW}$	(1.) _____
(2.) _____	(2.) Identity
(3.) △ YMW ≅ △ YNW	(3.) _____
(4.) _____	(4.) CPCTC
(5.) _____	(5.) If a ray separates an angle into 2 ≅ parts, that ray bisects the given angle.

Constructing Angles of Given Measures

4. a. Construct an angle whose measure is 60^0.

   b. Using the result from part (a.), construct an angle whose measure is 30^0.

5. a. By constructing perpendicular lines, construct an angle whose measure is 90^0.

   b. Using the result from part (a.), construct an angle whose measure is 45^0.

Regular Polygons

6. a. Use the formula $I = \dfrac{(n-2) \cdot 180^0}{n}$ to determine the measure of each interior angle of a regular hexagon. _____.

   b. Explain how to use the result found in 4(a) in order to construct an angle with the measure found in 6(a).

   _____

7. a. Use the formula $I = \dfrac{(n-2) \cdot 180^0}{n}$ to determine the measure of each interior angle of a regular octagon. _____.

   b. Explain how to use the result found in 5(b) in order to construct an angle with the measure found in 7(a).

   _____

Inequalities

1. Definition: Where a > b (read "a is greater than b"), there is a _____ number p
   for which a = b + p.

2. Because 7 > 5, we can determine a positive number that completes this statement; 7 = 5 + ___.

3. Because 9 = 2 + 7 (with 2 and 7 both positive), we know that 9 > ___ and 9 > ___.

Basic Geometric Inequalities (based upon lemmas)

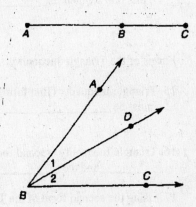

4. According to the Segment-Addition Postulate, AC = AB + BC.
   Then AC > ___ and AC > ___.

5. According to the Angle-Addition Postulate,
   m ∠ ABC = m ∠ 1 + m ∠ 2 in the figure.
   Then m ∠ ABC > ___ and m ∠ ABC > ___.

6. The measure of an exterior angle of a triangle is greater
   than the measure of either _____ interior angle.

7. In a right triangle, the largest angle is the _____ angle.

8. Complete (Addition Property of Inequality): If a > b and c > d,
   then _____.

Inequalities in a Triangle

9. Complete: If one side of a triangle is longer than a second side, then the measure of the angle
   opposite the longer side is _____ than the measure of the angle opposite the shorter side.

10. In △ ABC, it is given that AC > AB > BC.
    a. Which angle of △ ABC has the greatest measure? ___.
    b. Which angle of △ ABC has the smallest measure? ___.
    c. Write a three-part inequality that compares the measures of the
       angles of △ ABC. _____.

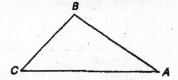

11. Complete: If the measure of one angle of a triangle is greater than the measure of a second angle,
    then the length of the side opposite the larger angle is _____ than the length of the side
    opposite the smaller angle.

12. Given the measures as shown in △ ABC,
    a. Find m ∠ C. _____.
    b. Name the longest side. _____.
    c. Name the shortest side. _____.
    d. Write a three-part inequality that compares the lengths of the
       sides of △ ABC. _____.

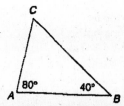

35

Further Inequalities

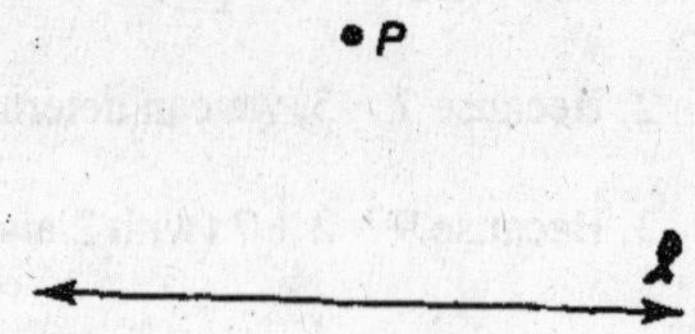

13. Describe the line segment whose length represents the shortest distance from point P to line ℓ ? _____.

•P

14. Compare to Exercise 13:
   "The shortest distance from a point not on a plane to the plane is the length of the line segment drawn from the point so that this line segment is _____ to the plane."

Forms of the Triangle Inequality

15. Triangle Inequality (first form): The sum of the lengths of any two sides of a triangle must be _____ _____ the length of the third side.

16. Triangle Inequality (second form): The length of any side of a triangle must lie between the _____ and the _____ of the lengths of the remaining sides.

17. Using the second form of the Triangle Inequality (see Exercise 16), can a triangle have these lengths for its three sides?
   a. 4, 5, and 6? _____   b. 4, 5, and 9? _____   c. 4, 5, and 11? _____.

Proof with Inequality Relationships

18. Complete the missing statements and reasons for the following proof.

   Given: AB > CD and BC > DE
   Prove: AC > CE

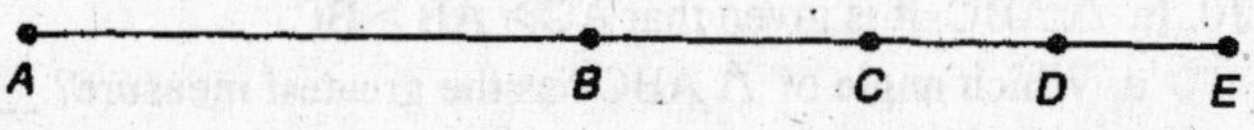

   A   B   C   D   E

Proof

Statements	Reasons
(1.) AB > CD and BC > DE	(1.) _____.
(2.) AB + BC > CD + DE	(2.) Addition Property of Inequality
(3.) AB + BC = AC and CD + DE = CE	(3.) _____.
(4.) _____	(4.) _____.

The Parallelogram

1. Complete: A parallelogram is a quadrilateral in which both pairs of opposite sides are _____.

2. Complete missing statements and reasons in the proof of the theorem:
   "A diagonal of a parallelogram separates it into two congruent triangles."

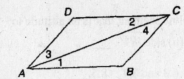

Given: ▱ ABCD with diagonal $\overline{AC}$
Prove: △ ACD ≅ △ CAB

Proof

Statements	Reasons
(1.) ▱ ABCD with diagonal $\overline{AC}$	(1.) _____
(2.) $\overline{AB} \parallel \overline{CD}$	(2.) By definition, the opposite sides of a ▱ are parallel.
(3.) ∠1 ≅ ∠2	(3.) If 2 ∥ lines are cut by a trans, the _____ _____ ∠s are ≅.
(4.) _____	(4.) Same as reason 2
(5.) _____	(5.) Same as reason 3
(6.) $\overline{AC} \cong \overline{AC}$	(6.) _____
(7.) △ ACD ≅ △ CAB	(7.) _____

Corollaries of the Preceding Theorem

3. Complete: Opposite angles of a parallelogram are _____.

4. Complete: Opposite sides of a parallelogram are _____.

5. Complete: The diagonals of a parallelogram _____ each other.

6. Complete: Consecutive angles of a parallelogram are _____.

Problems Based upon Corollaries

7. In ▱ ABCD, AD = 6.3 and AB = 8.7. Find: (a) DC ____ (b) BC ____.

8. In ▱ ABCD, m∠A = 72°. Find: (a) m∠C ____ (b) m∠B ____.

9. In ▱ ABCD, AD = x + 1 and AB = 2x - 5. If the perimeter of ▱ ABCD is 56, find: (a) x ____ (b) AD ____ (c) DC ____.

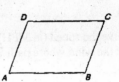

10. In ▱ ABCD, m∠A = y + 10 and m∠C = 3y - 90.
    Find: (a) y ____ (b) m∠A ____ (c) m∠B ____.

Exercises 7-10

37

11. In ▱ ABCD (not shown), m∠A = y + 10 and m∠B = 3y - 90.
    Find: (a) y _____ (b) m∠A _____ (c) m∠B _____.

Altitude of a Parallelogram

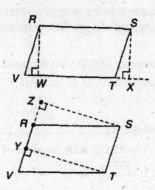

12. In ▱ RSTV, name a line segment that is an altitude to:
    (a) side $\overline{RS}$ _____ (b) side $\overline{RV}$ _____.

13. In ▱ RSTV , RW = 6.3 and YT = 8.5.
    Find the height of the parallelogram to:
    (a) side $\overline{RS}$ _____ (b) side $\overline{ST}$ _____.

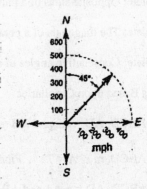

Unequal Diagonals of a Parallelogram

14. In ▱ MNPQ, it is given that m∠M > m∠N.
    Which diagonal ($\overline{QN}$ or $\overline{MP}$) is longer? _____.

15. In ▱ MNPQ, diagonal $\overline{QN}$ (if drawn) would be
    longer than diagonal $\overline{MP}$. Which angle (∠P or ∠Q)
    would be larger? _____.

Speed and Direction

16. Give the speed (in MPH) and direction (like N 20° W)
    of the airplane whose path is indicated.
    _____ .

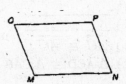

17. Give the speed (in MPH) and direction (like N 20° W)
    of the wind whose path is indicated.
    _____ .

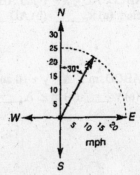

When a Quadrilateral is a Parallelogram

1. Complete this theorem: If two sides of a quadrilateral are both congruent and parallel, then the quadrilateral is a _____.

2. Complete: If both pairs of opposite sides of a quadrilateral are congruent, the quadrilateral is a _____.

3. Complete: If the diagonals of a quadrilateral _____ each other, the quadrilateral is a parallelogram.

4. Complete the missing statements and reasons for the proof of the theorem, "If two sides of a quadrilateral are both congruent and parallel, then the quadrilateral is a parallelogram."

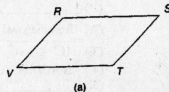

(a)

Given: $\overline{RS} \parallel \overline{VT}$ and $\overline{RS} \cong \overline{VT}$
Prove: RSTV is a parallelogram

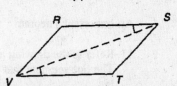

Proof

Statements	Reasons
(1.) _____	(1.) Given
(2.) Draw diagonal $\overline{VS}$	(2.) Through 2 points, there is one line.
(3.) $\overline{VS} \cong \overline{VS}$	(3.) _____
(4.) $\angle RSV \cong \angle TVS$	(4.) If 2 $\parallel$ lines are cut by a trans., the _____ _____ angles are $\cong$.
(5.) $\triangle RSV \cong \triangle TVS$	(5.) _____
(6.) $\therefore \angle RVS \cong$ _____	(6.) CPCTC
(7.) $\overline{RV} \parallel \overline{ST}$	(7.) If two lines are cut by a trans. so that alt. int. $\angle$s are $\cong$, then these lines are parallel.
(8.) _____	(8.) If both pairs of opposite sides of a quad. are parallel, the quadrilateral is a parallelogram.

The Kite

5. Complete this definition: A kite is a quadrilateral with two distinct pairs of congruent _____ sides.

6. Complete this theorem: In a kite, one pair of opposite _____ are congruent.

7. In kite ABCD, which pair of angles are congruent? _____.

8. If drawn, how would the diagonals $\overline{AC}$ and $\overline{BD}$ of kite ABCD be related? _____.

9. In kite ABCD, $\angle B$ is a right angle. Given that AB = 8 and BC = 6, find the length of diagonal $\overline{AC}$. _____.

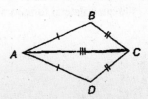

39

10. Complete the missing statements and reasons for the proof of the theorem,
    "In a kite, one pair of opposite angles are congruent."

    Given: Kite ABCD; $\overline{BC} \cong \overline{CD}$ and $\overline{AD} \cong \overline{AB}$
    Prove: $\angle B \cong \angle D$

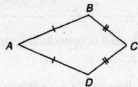
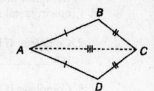

Proof

Statements	Reasons
(1.) _____	(1.) Given
(2.) Draw diagonal $\overline{AC}$	(2.) Through 2 points, there is one line.
(3.) $\overline{AC} \cong \overline{AC}$	(3.) _____
(4.) $\triangle ACD \cong \triangle ACB$	(4.) _____
(5.) _____	(5.) _____

An Important Theorem

11. Refer to the drawing at the right and complete this theorem:
    The line segment that joins the midpoints of two sides of a triangle
    is _____ to the third side and has a length equal
    to _____ the length of the third side.

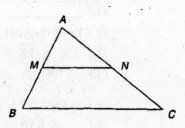

12. In $\triangle$ ABC, M and N are the midpoints of $\overline{AB}$ and $\overline{AC}$ respectively.
    (a.) If BC = 12.8, find MN. _____.
    (b.) If MN = 5.3, find BC. _____.
    (c.) If MN = x + 3, find an expression for BC. _____.
    (d.) If MN = x + 3 and BC = 4x – 12, find x. _____.

13. In $\triangle$ RST, points X, Y, and Z are the midpoints of the indicated sides.
    If RS = 18, RT = 24, and ST = 26, find:
    (a.) XY _____   (b.) YZ _____   (c.) XZ _____.
    (d.) the perimeter of $\triangle$ XYZ _____.

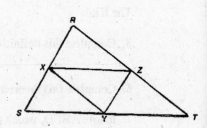

14. In $\triangle$ RST, points X, Y, and Z are the midpoints of the indicated sides.
    If XY = 7.2, XZ = 6.9, and YZ = 5.1, find:
    (a.) RS _____   (b.) RT _____   (c.) ST _____.
    (d.) the perimeter of $\triangle$ RST _____.

15. When the midpoints of the sides of any quadrilateral are joined in order,
    the quadrilateral formed will always be a _____.

40

The Rectangle

1. Complete: A rectangle is a _____ that has a right angle.

2. In rectangle ABCD, $\angle$ A is a right angle. Because a rectangle is a type of parallelogram, its opposite angles are congruent and m $\angle$ C = _____.

3. In rectangle ABCD, let m $\angle$ B = m $\angle$ D = x. Knowing that the sum of measures of interior angles of a quadrilateral is 360^0, it follows that x + x + 90 + 90 = 360. In turn, it follows that "All angles of a rectangle are _____ angles."

4. Complete missing statements and reasons in the proof of the theorem: "The diagonals of a rectangle are congruent."

   Given: Rectangle MNPQ with diagonals $\overline{MP}$ and $\overline{NQ}$

   Prove: $\overline{MP} \cong \overline{NQ}$

   ### Proof

Statements	Reasons
(1.) _____	(1.) _____
(2.) MNPQ is a parallelogram	(2.) By definition, a rectangle is a ▱ .
(3.) $\overline{MN} \cong \overline{PQ}$	(3.) _____ .
(4.) $\overline{MQ} \cong \overline{MQ}$	(4.) _____
(5.) $\angle$ s NMQ and PQM are rt. $\angle$ s	(5.) _____
(6.) _____	(6.) All right angles are congruent.
(7.) $\triangle$ NMQ $\cong$ $\triangle$ PQM	(7.) _____ .
(8.) _____	(8.) _____ .

The Square

5. Complete (Definition): A square is a rectangle that has two congruent _____ sides.

6. Because a square is a rectangle (and a type of parallelogram), it follows that "All four sides of a square are _____."

7. Like a rectangle, the diagonals of a square are _____.
   Unlike a rectangle, the diagonals of a square are also _____.

The Rhombus

8. Complete: A rhombus is a _____ with two congruent adjacent sides.

9. Because a rhombus is a type of parallelogram, the four sides of a rhombus must be _____.

10. Find the perimeter of rhombus ABCD if AB = 5.7 cm. _____

41

11. Complete the missing statements and reasons for the proof of the theorem,
"The diagonals of a rhombus are perpendicular."

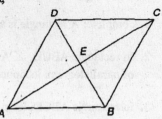

Given: Rhombus ABCD with diagonals $\overline{AC}$ and $\overline{BD}$

Prove: $\overline{AC} \perp \overline{BD}$

Proof

Statements	Reasons
(1.) _____	(1.) _____ :
(2.) ABCD is a ▱	(2.) By def'n, a rhombus is a _____
(3.) Diagonal $\overline{DB}$ bisects $\overline{AC}$	(3.) The diagonals of a ▱ _____ each other.
(4.) $\overline{AE} \cong \overline{EC}$	(4.) _____ .
(5.) $\overline{AD} \cong \overline{DC}$	(5.) All sides of a rhombus are _____ .
(6.) _____	(6.) Identity
(7.) △ ADE ≅ △ CDE	(7.) _____ .
(8.) ∠ DEA ≅ ∠ DEC	(8.) _____ .
(9.) _____	(9.) If 2 lines meet to form ≅ adj. ∠ s, then these lines are perpendicular.

The Pythagorean Theorem

12. With squares upon each of the sides of the triangle, the figure
at the right is a visual reminder of the Pythagorean Theorem,
"In a right triangle with hypotenuse length c and legs of
lengths a and b, it follows that _____."

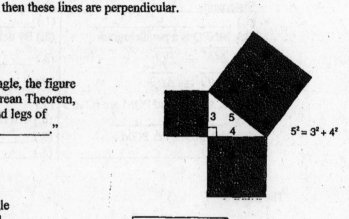

13. What is the length of the diagonal for a rectangle
whose adjacent sides have lengths of 3 feet and
4 feet? _____ .

14. What is the length of each side of the rhombus
whose diagonals measure 10 centimeters and
24 centimeters? _____ .

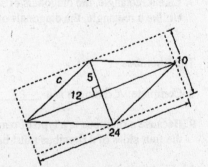

42

The Trapezoid

1. Definition: A trapezoid is a quadrilateral with exactly two sides that are _____.

2. Given trapezoid HJKL with $\overline{HL} \parallel \overline{JK}$, name:
   (a.) the bases _____ (b.) the legs _____.
   (c.) one pair of base angles _____.

3. Given trapezoid HJKL with $\overline{HL} \parallel \overline{JK}$, is HJKL an isosceles
   trapezoid if: (a.) $\overline{HJ} \cong \overline{HL}$? _____ (b.) $\overline{HJ} \cong \overline{LK}$? _____.

4. Given trapezoid HJKL with $\overline{HL} \parallel \overline{JK}$ and midpoints M and N as shown,
   line segment $\overline{MN}$ is the _____ of trapezoid HJKL.

5. Intuition suggests that $\overline{MN}$ is _____ to each base.

6. In a trapezoid, a line segment from one vertex of a trapezoid that is
   perpendicular to the opposite base is a(n) _____ of the trapezoid.

Theorems about Trapezoids

7. Complete: The base angles of an isosceles trapezoid are _____.

8. Complete: The diagonals of a(n) _____ trapezoid are congruent.

9. Complete: The length of the median of a trapezoid is equal to
   one-half the sum of the lengths of the two _____.

Problems involving Trapezoids

10. In trapezoid HJKL with median $\overline{MN}$, suppose that m$\angle$H = 123^0,
    m$\angle$K = 72^0, HL = 14, and JK = 22. Find:
    (a.) m$\angle$J _____ (b.) m$\angle$L _____ (c.) MN _____.

11. In trapezoid HJKL with median $\overline{MN}$, suppose that HL = 15
    and MN = 23. Find JK. _____.

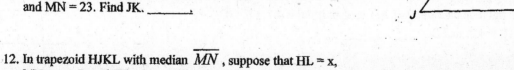

12. In trapezoid HJKL with median $\overline{MN}$, suppose that HL = x,
    MN = x + 7, and JK = 3x - 6. Find:
    (a.) x _____ (b.) HL _____ (c.) MN _____ (d.) JK _____.

When a Trapezoid is Isosceles

13. Complete: If a pair of base angles of a trapezoid are _____, then the trapezoid is isosceles.

14. Complete: If the diagonals of a trapezoid are congruent, then the trapezoid is _____.

Proof of a Theorem

15. Complete the missing statements and reasons for the proof of the following theorem:
    "The base angles of an isosceles trapezoid are congruent."

Given: Trapezoid RSTV with $\overline{RS} \parallel \overline{VT}$
and $\overline{RV} \cong \overline{ST}$
Prove: $\angle V \cong \angle T$

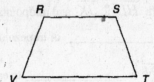 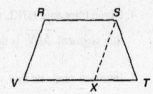

Proof

Statements	Reasons
(1.) _____	(1.) Given
(2.) Draw $\overline{SX} \parallel \overline{RV}$ as shown	(2.) Parallel Postulate
(3.) RSXV is a ▱	(3.) _____
(4.) $\overline{RV} \cong \overline{SX}$	(4.) Opposite sides of a ▱ are _____.
(5.) $\overline{ST} \cong \overline{SX}$	(5.) Transitive Property of Congruence
(6.) $\angle SXT \cong \angle T$	(6.) If 2 sides of a Δ are ≅, then the angles _____ these sides are ≅.
(7.) $\angle V \cong \angle SXT$	(7.) If 2 ∥ lines are cut by a trans, then the _____ angles are ≅.
(8.) _____	(8.) _____

One More Property

16. Consider the theorem, "If 3 (or more) parallel lines intercept congruent segments on one transversal, then they intercept congruent segments on any transversal."

    In the figure at the right, a ∥ b ∥ c; also, $\overline{DE} \cong \overline{EF}$.
    Using the theorem, what conclusion can you draw? _____

17. In the figure at the right, a ∥ b ∥ c; also, if DE ≈ 4.7, EF = 4.7, and AB = 4.2, find BC. _____

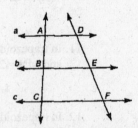

44

Ratios and Rates

1. Write each ratio in lowest terms. Units are not necessary in the answer.
   (a.) 6 to 8 _____   (b.) 12 cm : 20 cm _____   (c.) 4 inches : 1 foot _____.
   [Hint: 1 foot = 12 inches.]

2. Write each rate in simplest form. Units are necessary in the answer.
   (a.) 240 miles : 15 gallons _____   (b.) 56 cents : 4 pencils _____ .

Proportions

3. In the proportion $\frac{a}{b} = \frac{c}{d}$, a and d are known as the _____ of the proportion while b and c are the _____ of the proportion.

4. Solve each proportion for x:
   (a.) $\frac{x}{5} = \frac{7}{9}$ _____   (b.) $\frac{x+1}{2} = \frac{4}{x-1}$ _____.

   (c.) $\frac{x}{4} = \frac{x-2}{3}$ _____   (d.) $\frac{x+3}{5} = \frac{4}{x-5}$ _____.

5. When the second and third terms of a proportion are alike, as in $\frac{a}{b} = \frac{b}{c}$, the number b is known as the _____ _____ of the proportion.

6. Find the geometric mean(s) for 4 and 9. _____.

Applications

7. If an automobile can travel 145 miles using 5 gallons of gasoline, how far can it travel when consuming 7 gallons of gasoline? _____.

8. If two unknown numbers are in the ratio a : b, then the numbers can be represented by ax and bx. Use the preceding fact to find the measures of two complementary angles that are in the ratio 2 : 3. _____ and _____.

9. When three unknown numbers are in the ratio a : b : c, these numbers can be represented by ax, bx, and cx. Use the preceding fact to find the measures of the three angles of a triangle whose measures are in the ratio 1 : 2 : 3. _____, _____, and _____.

Properties of a Proportion

10. One of the claims found in Property 2 of this section states that "the means of a true proportion can be interchanged to form another true proportion." By using the preceding property, replace the true proportion $\frac{12}{18} = \frac{2}{3}$ by another true proportion. _____.

11. One of the claims found in Property 3 of this section takes the algebraic form, "If $\frac{a}{b} = \frac{c}{d}$, then $\frac{a+b}{b} = \frac{c+d}{d}$." Given the proportion, $\frac{6}{8} = \frac{3}{4}$, use this property to write another true proportion. _____.

Extended Proportions

12. An extended proportion takes the form $\frac{a}{b} = \frac{c}{d} = \frac{e}{f} = \ldots$. Determine the number that should replace the question mark in the following extended proportion:

$$\frac{2eggs}{3cups} = \frac{4eggs}{6cups} = \frac{?eggs}{9cups}$$ _____.

13. In $\triangle ABC$ and $\triangle DEF$, it is known that $\frac{AB}{DE} = \frac{AC}{DF} = \frac{BC}{EF}$.

Using the lengths provided in the figure, find x (length of $\overline{DF}$) and y (length of $\overline{EF}$). x = _____ and y = _____.

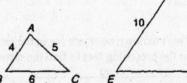

Similar Polygons

1. (a.) True or False: Any two congruent polygons are also similar polygons. _____.
   (b.) True or False: Any two similar polygons are also congruent polygons. _____.

2. Given that quadrilateral ABCD and quadrilateral RSTV are similar,
   (a.) which angle of quad. RSTV corresponds to $\angle$ A of quad. ABCD? _____.
   (b.) which side of quad. ABCD corresponds to side $\overline{TV}$ of quad. RSTV? _____.

3. By definition, two polygons are *similar* if:
   (a.) All pairs of corresponding angles are _____.
   (b.) All pairs of corresponding sides are _____.

4. Which figures must be similar?
   (a.) any two isosceles triangles? _____   (b.) any two equilateral triangles? _____.
   (c.) any two squares? _____   (d.) any two rectangles? _____.

Problems involving Similar Polygons

5. Given that $\triangle$ ABC ~ $\triangle$ XTN with m $\angle$ A = 92^0 and m $\angle$ T = 27^0, find:
   (a.) m $\angle$ B _____   (b.) m $\angle$ X _____   (c.) m $\angle$ C _____.

6. Given that $\triangle$ ABC ~ $\triangle$ XTN, find the lengths x and y that

   complete the extended proportion: $\dfrac{AB}{XT} = \dfrac{x}{TN} = \dfrac{AC}{y}$.

   (a.) x _____   (b.) y _____.

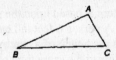
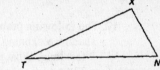

7. Given that $\triangle$ ABC ~ $\triangle$ XTN, suppose that AB = 7, AC = 4,   Exercises 5-8
   BC = 8, and XT = 10. Find:
   (a.) XN _____   (b.) TN _____.

8. When polygons are similar, the lengths of corresponding sides have a constant of proportionality.

   Where $\triangle$ ABC ~ $\triangle$ XTN, suppose that AB = 6 and XT = 9. Then AB = $\dfrac{2}{3}$ (XT) and it follows

   that AC = $\dfrac{2}{3}$ (XN). If XN = 6.9, find AC. _____.

Further Problems

9. Suppose that quadrilateral ABCD ~ quadrilateral HJKL, with congruent angles marked as shown.
   If m∠A = x, m∠C = 2x, and m∠K = 3(x - 10), find:
   (a.) x _____   (b.) m∠K _____.

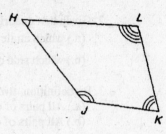

10. Suppose that quadrilateral ABCD ~ quadrilateral HJKL,
    with congruent angles marked as shown. If AB = 6, BC = 4,
    CD = 5, DA = 7, and HJ = 12, find the perimeter of
    quadrilateral HJKL. _____.

Exercises 9 & 10

Selected Applications

11. On a blueprint, the length of a rectangular room that is actually 18 feet long is represented by a
    line segment that is 3.6 inches in length. By what length (on the blueprint) is the 15 foot width of the
    room represented? _____.

12. The following problem is based upon the concept known as
    shadow reckoning. A person 6 feet tall casts a shadow that is
    9 feet long. What is the height of a nearby flagpole that casts
    a 30 foot long shadow at the same time of day? _____.

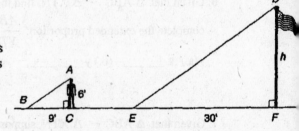

13. It is given that Δ ADE ~ Δ ABC. If DE = 3, AC = 16,
    and EC = BC, find the length BC. _____.
    [Hint: There are 2 solutions to this problem.]

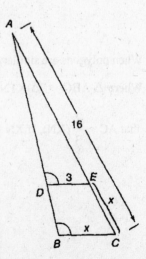

48

A Method of Proving that Triangles are Similar

1. While AAA (Postulate 15) can be used to prove that 2 triangles are similar, it is better to apply the corollary AA because it involves showing that only _____ pairs of angles are congruent.

2. Are the triangles similar if it is known that:

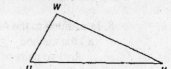

   (a.) $\angle T \cong \angle W$ and $\angle R \cong \angle U$? _____
   (b.) $\angle T \cong \angle W$, $m \angle R = 65^0$ and $m \angle S = 25^0$? _____
   (c.) $\angle T \cong \angle R$ and $\angle W \cong \angle U$? _____

Proof Using AA

3. Complete missing statements and reasons in the proof of the theorem:

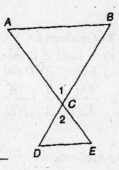

   Given: $\overline{AB} \parallel \overline{DE}$
   Prove: $\triangle ABC \sim \triangle EDC$

   Proof

Statements	Reasons
(1.) _____	(1.) _____
(2.) $\angle A \cong \angle E$	(2.) If 2 $\parallel$ lines are cut by a trans, _____
(3.) $\angle 1 \cong \angle 2$	(3.) _____
(4.) _____	(4.) _____

4. Once triangles have been proven similar, we can also show that:
   (a.) corresponding _____ are proportional by the reason CSSTP; and
   (b.) corresponding _____ are congruent by the reason CASTC.

5. Complete missing statements and reasons in the proof of the theorem:

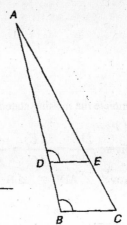

   Given: $\angle ADE \cong \angle B$
   Prove: $\dfrac{DE}{BC} = \dfrac{AE}{AC}$

   Proof

Statements	Reasons
(1.) _____	(1.) _____
(2.) $\angle A \cong \angle A$	(2.) _____
(3.) $\triangle ADE \sim \triangle ABC$	(3.) _____
(4.) _____	(4.) CSSTP

6. In problem 5, suppose that you had been asked to prove that $DE \cdot AC = BC \cdot AE$. That would add a step 5 to the proof shown. The reason for the last statement of the proof is known as the _____-_____ Product Property.

7. Complete this theorem: The lengths of two corresponding altitudes of two similar triangles have the same ratio as the lengths of any two corresponding _____. .

Further Methods for Establishing that Two Triangles are Similar

8. In addition to the AA method for showing that two triangles are similar,
   (a.) the method _____ is used when a pair of congruent included angles are formed by two pairs of sides that are proportional in length; and
   (b.) the method _____ is used when the three pairs of sides are proportional in length.

9. As a point of clarification, SAS and SSS are used to prove that two triangles are _____ while SAS ~ and SSS ~ are used to prove that two triangles are _____.

10. Which method (AA, SAS ~ , or SSS ~) proves that $\triangle$ HJK is similar to $\triangle$ FGK if:
   (a.) $\angle$ J and $\angle$ G are right angles? _____.
   (b.) KG = 2.7(KJ) and FK = 2.7(HK)? _____.
   (c.) $\dfrac{KG}{KJ} = \dfrac{3}{2}$, $\dfrac{FK}{HK} = \dfrac{3}{2}$, and $\dfrac{FG}{HJ} = \dfrac{3}{2}$? _____.

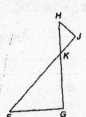

Further Applications

11. In the figure, $\dfrac{DG}{DE} = \dfrac{DH}{DF}$. If m$\angle$E = x, m$\angle$DGH = 2x − 80, and m$\angle$F = x − 45, find:
   (a.) m$\angle$E _____ (b.) m$\angle$F _____ (c.) m$\angle$D _____ .

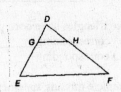

12. Complete the missing statements and reasons for the following proof.

Given: $\dfrac{DE}{BC} = \dfrac{AE}{AC} = \dfrac{AD}{AB}$

Prove: $\angle$ADE $\cong$ $\angle$B

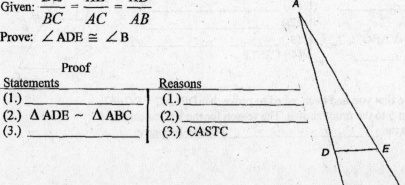

Proof

Statements	Reasons
(1.) _____	(1.) _____
(2.) $\triangle$ ADE ~ $\triangle$ ABC	(2.) _____
(3.) _____	(3.) CASTC

50

An Important Theorem

1. Complete: The altitude drawn to the hypotenuse of a right triangle separates the triangle into two right triangles that are _____ to each other and to the given right triangle.

2. Consider the theorem above and the figures below. To prove that $\triangle ADC \sim \triangle ACB$, we use the method AA. By the reason _____, we know that $\angle A \cong \angle A$. Also, $\angle D$ of $\triangle ADC$ and $\angle C$ of $\triangle ACB$ are both _____ angles, so these angles are congruent as well.

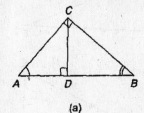

(a)

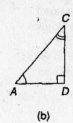

(b)

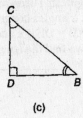

(c)

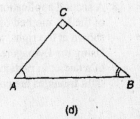

(d)

Proof of the Pythagorean Theorem

3. Consider the figures at the right. From the fact that $\triangle ABC \sim \triangle CBD$,

we have $\dfrac{c}{a} = \dfrac{a}{y}$ and that $a^2 =$ _____ . Because $\triangle ABC \sim \triangle ACD$,

we also know that $\dfrac{c}{b} = \dfrac{b}{x}$ and that $b^2 =$ _____ . By substitution, it

follows that $a^2 + b^2 = cy + cx = c(y + x)$. Applying the Segment-Addition Postulate in figure (b), we see that $y + x =$ _____. Thus, the preceding equation can be replaced by $a^2 + b^2 = c \cdot c$ or $a^2 + b^2 = c^2$.

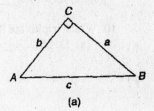

(a)

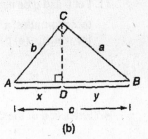

(b)

4. In right $\triangle ABC$ with hypotenuse length c and lengths of legs a and b, find:
   (a.) c if a = 4 and b = 6 _____   (b.) b if a = 12 and c = 13 _____.

The Converse of the Pythagorean Theorem

5. Complete. Let a, b, and c represent the lengths of the sides of a triangle, with c being the length of the longest side. If $c^2 = a^2 + b^2$, then the triangle is a _____ triangle with its right angle being opposite the side of length _____.

6. Given these lengths for the 3 sides of a triangle, is the triangle a right triangle?
   (a.) a = 5, b = 12, and c = 13? _____   (b) a = 7, b = 9, and c = 10? _____.
   (c.) a = $\sqrt{2}$, b = $\sqrt{3}$, and c = $\sqrt{5}$? _____.

Applications of the Pythagorean Theorem

7. A strong wind holds a kite 30 feet above the ground in a position
   40 feet across the ground. With the string to the kite along a line,
   how much string does the girl have out to the kite? _____

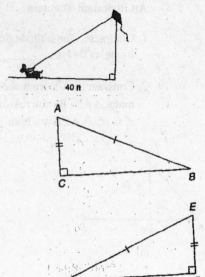

8. A second application of the Pythagorean Theorem allows for the proof
   of the HL method of proving right triangles congruent. For the triangles
   shown at the right, we know that AB = DE = c and that AC = EF = a.
   Using the Pythagorean Theorem, we find the lengths CB and DF in terms
   of a and c; in particular, CB = DF = _____. By SSS, the two
   right triangles are congruent.

Pythagorean Triples

9. Complete this definition: A Pythagorean Triple is a set of three natural numbers
   (a,b,c) for which _____.

10. Which triples are Pythagorean Triples?
    (a) (3,4,5)? _____   (b) (4,5,6)? _____   (c) (0.3,0.4,0.5)? _____   (d) (20,21,29)? _____

11. Let p and q be natural numbers with p > q. Then (a,b,c) is a Pythagorean Triple generated by
    these formulas: $a = p^2 - q^2$, $b = 2pq$, and $c = p^2 + q^2$. Find the Pythagorean Triples (a,b,c)
    for which: (a) p = 2 and q = 1 _____   (b) p = 3 and q = 2 _____

An Extension of the Converse

12. Complete: Let a, b, and c represent the lengths of the sides of a triangle, with c being the
    length of the longest side.
    (a) If $c^2 = a^2 + b^2$, the triangle is a(n) _____ triangle.
    (b) If $c^2 < a^2 + b^2$, the triangle is a(n) _____ triangle.
    (c) If $c^2 > a^2 + b^2$, the triangle is a(n) _____ triangle.

13. What type of triangle (acute, right, or obtuse) has these lengths for its sides?
    (a) a = 3, b = 4, and c = 6? _____
    (b) a = 3, b = 4, and c = 5? _____
    (c) a = 4, b = 5, and c = 6? _____

52

The 45°-45°-90° Right Triangle

1. In any triangle with 2 congruent angles, the sides opposite these angles are also _____

   Thus, the sides opposite the 45° angles of a 45°-45°-90° right triangle are _____

   If the length of the side opposite one 45° angle measures a, then the length

   of the side opposite the second 45° angle must measure _____

2. If both legs of a 45°-45°-90° right triangle have length a,
   then the length c of the hypotenuse can be found by solving the

   equation $c^2 = a^2 + a^2$. In terms of a, we find that c = _____

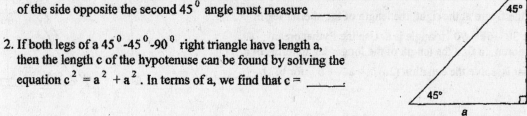

3. Complete the 45°-45°-90° Theorem: In a triangle whose angle measures are 45°, 45°, and 90°,
   the hypotenuse has a length equal to the product of ____ and the length of either of the congruent legs.

For Exercises 4 and 5, apply the 45°-45°-90° Theorem.

4. Given: Δ ABC is the 45°-45°-90° triangle shown.
   (a.) If AC = 6, find BC ____ and AB ____.

   (b.) If AB = $8\sqrt{2}$, find AC ____ and BC ____.

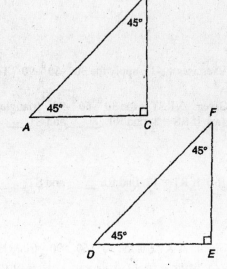

5. Given: Δ DEF is the 45°-45°-90° triangle shown.
   (a.) If EF = $3\sqrt{2}$, find DE ____ and DF ____.

   (b.) If DF = 4, find DE ____ and EF ____.

6. The figure at the right is a geometric representation
   of the _____ Theorem.

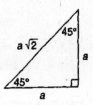

7. Repeat problem 4(a), but find the length of hypotenuse $\overline{AB}$ by using the Pythagorean Theorem
   and the fact that AC = 6 and BC = 6. _____

The 30^0-60^0-90^0 Right Triangle

8. Use the figure provided. If the length of the shorter leg of the 30^0-60^0-90^0 triangle is a, then the hypotenuse must have the length _____.

9. In the figure at the right, the length of the shorter leg of the 30^0-60^0-90^0 triangle is a. Use the Pythagorean Theorem to find the length of the longer leg in terms of a. That is, solve the equation $(2a)^2 = a^2 + b^2$ for b. b = _____

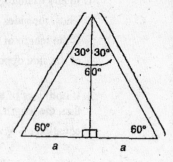

10. Complete the 30^0-60^0-90^0 Theorem: In a triangle whose angle measures are 30^0, 60^0, and 90^0, the hypotenuse has a length equal to twice the length of the _____ leg while the length of the longer leg is the product of _____ and the length of the shorter leg.

For Exercises 11-13, apply the 30^0-60^0-90^0 Theorem.

11. Given: $\triangle$ RST is the 30^0-60^0-90^0 triangle shown.
    (a.) If RS = 8, find RT _____ and ST _____.

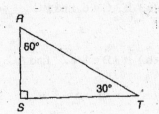

    (b.) If RT = 12, find RS _____ and ST _____.

12. Given: $\triangle$ XYZ is the 30^0-60^0-90^0 triangle shown.
    (a.) If YZ = $10\sqrt{3}$, find XY _____ and XZ _____.

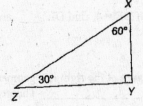

    (b.) If XY = $2\sqrt{3}$, find YZ _____ and XZ _____.

13. The figure at the right is a geometric representation of the _____ Theorem.

Segments Divided Proportionally

1. In the figure, $\overline{AC}$ and $\overline{DF}$ are divided proportionally at B and E if $\dfrac{AB}{DE} = \dfrac{BC}{EF}$ or if $\dfrac{AB}{BC}$ equals _____.

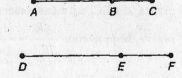

2. In the figure, $\overline{AC}$ and $\overline{DF}$ are divided proportionally at B and E. If AB = 5, BC = 3, and DE = 7, find EF. _____

3. Complete this theorem (see figure below): If a line is parallel to one side of a triangle and intersects the other two sides, it divided these sides _____.

4. A sketch of the proof for the theorem in problem 3:

   Given: Δ ABC with $\overline{DE} \parallel \overline{BC}$

   Prove: $\dfrac{DB}{AD} = \dfrac{EC}{AE}$

   Proof: Because $\overline{DE} \parallel \overline{BC}$, corresponding angles are congruent; that is, $\angle 1 \cong \angle 2$. Also, $\angle A \cong \angle A$ by the reason _____. Then Δ ADE $\sim \Delta$ ____ by AA.

   By CSSTP, it follows that $\dfrac{AB}{AD} = \dfrac{AC}{AE}$ (*). Using a property

   of proportions, it follows that $\dfrac{AB - AD}{AD} = \dfrac{AC - AE}{AE}$.

   But AB – AD = ____ and AC – AE = ____, so the preceding proportion becomes _____.

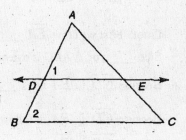

Exercises 3-6

5. Given that $\overline{DE} \parallel \overline{BC}$, AD = 6, DB = 4, and EC = 5, find AE. _____

6. Given that $\overline{DE} \parallel \overline{BC}$, AD = x + 2, DB = x, AE = x + 4, and EC = x + 1, find x. _____

7. Complete: When three (or more) parallel lines are cut by a transversal, the transversals are divided proportionally by the _____ lines.

8. In the figure, $\ell_1 \parallel \ell_2 \parallel \ell_3$, AC = 12, BC = 7, and DE = 6. Find EF. _____

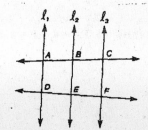

The Angle-Bisector Theorem

9. In $\triangle ABC$, $AB = 4$, $BC = 6$, and $AC = 5$. Given that $\overrightarrow{BD}$ bisects $\angle ABC$, use measurements shown in the drawing to rewrite the proportion $\dfrac{AB}{BC} = \dfrac{AD}{DC}$. _____

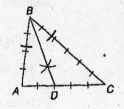 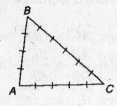

10. Complete this theorem: If a ray bisects an angle of a triangle, then it divides the opposite side into line segments whose lengths are proportional to the lengths of the two sides that form the _____ angle.

11. A sketch of the proof for the theorem in problem 10:

Given: $\triangle ABC$ in which $\overrightarrow{CD}$ bisects $\angle ACB$.

Prove: $\dfrac{AD}{AC} = \dfrac{DB}{CB}$

Proof: First we draw $\overline{EA}$ _____ to $\overline{CD}$.

Side _____ of $\triangle ABC$ is extended to meet $\overrightarrow{EA}$.

Because $\overline{CD} \parallel \overline{EA}$ in $\triangle$ _____, we have $\dfrac{EC}{AD} = \dfrac{CB}{DB}$ (*).

Because $\overline{CD} \parallel \overline{EA}$, it follows that $\angle 2 \cong \angle$ _____ (alt. int. $\angle$s).

Because $\overline{CD} \parallel \overline{EA}$, it follows that $\angle 1 \cong \angle$ _____ (corr. $\angle$s).

But $\angle 1 \cong \angle 2$ (bisected angle), so $\angle 3 \cong \angle 4$ by the Transitive Property of Congruence.

Then $\overline{AC} \cong \overline{EC}$ because they lie opposite congruent angles 3 and 4 of $\triangle$ _____.
With $AC = EC$, the starred (*) proportion becomes _____ or _____ by inverting each ratio of this equation.

12. In $\triangle XYZ$, $\overrightarrow{YW}$ bisects $\angle XYZ$. If $XY = 4$, $YZ = 6$, and $XW = 3$, find WZ. _____

13. In $\triangle XYZ$, $\overrightarrow{YW}$ bisects $\angle XYZ$. If $XY = 6$, $YZ = 8$, and $XZ = 7$, find WZ. _____

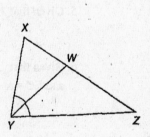

Ceva's Theorem

14. According to Ceva's Theorem, $\dfrac{BF}{FA} \cdot \dfrac{AE}{EC} \cdot \dfrac{CD}{DB} = 1$.

If $BF = 5$, $FA = 4$, $AE = 3$, and $EC = 4$, find the ratio $\dfrac{CD}{DB}$. _____

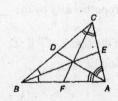

56

Basic Terminology for Circles

1. In ⊙ Q, $\overline{QS}$ is a _____ while $\overline{QW}$, $\overline{QV}$, and $\overline{QT}$ are _____ (plural of radius).
$\overline{SW}$ is a _____ of ⊙ Q while $\overline{WT}$ is a chord and
also a(n) _____ because it contains the center Q.

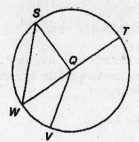

2. In ⊙ Q, suppose that $\overline{SQ} \perp \overline{WT}$. If QV = 5, find:
   (a) QS ____ (b) WT ____ (c) SW ____.

3. Congruent circles have the same length of _____.
   Concentric circles have the same _____.

Central Angles

4. In ⊙ O, $\overline{MP}$ and $\overline{QN}$ are diameters. Because the vertex of ∠ 1 is the center O
   of the circle and the sides of ∠ 1 are radii, ∠ 1 is a(n) _____ angle. Also,
   $\overparen{NP}$ is a(n) _____ arc while $\overparen{NPQM}$ is a (n) _____ arc
   and $\overparen{NPQ}$ is a(n) _____.

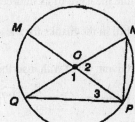

5. Complete (Central Angle Postulate): In a circle, the measure of a
   central angle is equal to the measure of its intercepted _____.

6. In ⊙ O, $\overline{MP}$ and $\overline{QN}$ are diameters. If m $\overparen{NP}$ = 58°, find:
   (a) m ∠ 2 ____ (b) m ∠ 1 ____ (c) m $\overparen{QP}$ ____ (d) m ∠ 3 ____

Further Properties

7. Fill the blanks in the paragraph proof of this theorem:
   "A radius that is perpendicular to a chord bisects the chord."

   Given: In ⊙ O, radius $\overline{OD} \perp \overline{AB}$ at point C.
   Prove: $\overline{OD}$ bisects $\overline{AB}$

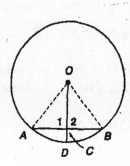

   Proof: In ⊙ O, it is "given" that _____.
   We draw auxiliary radii $\overline{OA}$ and _____ as shown. Now
   $\overline{OA} \cong \overline{OB}$ because "All radii of a circle are _____."
   With right ∠ 1 ≅ right ∠ 2 and _____ by Identity,
   it follows that Δ OCA ≅ Δ _____ by HL. Then $\overline{AC} \cong \overline{CB}$
   by CPCTC. By definition, $\overline{OD}$ bisects $\overline{AB}$.

8. In ⊙O, radius $\overline{OD} \perp$ chord $\overline{AB}$ at C. If AB = 8 and OB = 5, find OC. _____.

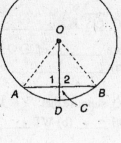

9. In ⊙O, m$\overarc{AD}$ + m$\overarc{DB}$ = _____.

10. For two arcs to be congruent, the arcs must have the same measure and the arcs must be arcs of a circle or be arcs of _____ circles.

Inscribed Angles

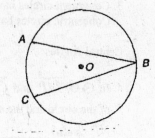

11. In ⊙O, $\overline{AB}$ and $\overline{BC}$ are _____. With the vertex of ∠B on ⊙O, ∠B is known as a(n) _____ angle. For ∠B, the intercepted arc is _____.

12. Complete: The measure of an inscribed angle equals _____ of the measure of its intercepted arc.

13. Fill in the blanks for the proof of Case 1 of the theorem in Exercise 12.

Given: ⊙O with inscribed ∠RST and diameter $\overline{TS}$

Prove: m∠S = $\dfrac{1}{2}$ m$\overarc{RT}$

Proof: ⊙O with inscribed ∠RST and diameter $\overline{TS}$. Draw radius _____. Because ∠ROT is a central angle, m∠ROT = m$\overarc{RT}$. With $\overline{OR} \cong \overline{OS}$, it follows that m∠R = m∠____ in △ROS. Because ∠ROT is an exterior angle of △ROS, it follows that m∠ROT = _____ + _____. By substitution, we have m∠ROT = m∠S + m∠S and m∠ROT = 2(m∠S). Then 2(m∠S) = m$\overarc{RT}$; in turn, m∠S = $\dfrac{1}{2}$ m$\overarc{RT}$.

14. In ⊙O, m$\overarc{AC}$ = 54^{0} and m$\overarc{ED}$ = 38^{0}. Find:
   (a) m∠B _____   (b) m∠O _____.

15. In ⊙O, m∠B = 29^{0} and m∠O = 36.7^{0}. Find:
   (a) m$\overarc{AC}$ _____   (b) m$\overarc{ED}$ _____.

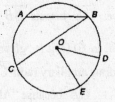

Further Properties

16. Complete (See figure at right): An angle inscribed in a semicircle is a(n) _____ angle.

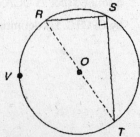

17. Complete: If two inscribed angles intercept the same arc, these angles are _____.

58

Tangents and Secants

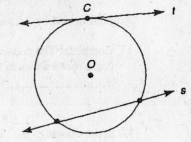

1. Because line t intersects (touches) ⊙ O at one point, line t
   is a(n) _____ to ⊙ O. Point C is known as the
   point of contact or as the point of _____.

2. Because line s intersects ⊙ O in two points, line s
   is a(n) _____ to ⊙ O.

Inscribed and Circumscribed Polygons and Circles

3. A polygon (like quadrilateral RSTV) is _____ in a circle
   if its sides are _____ of the circle. Because quad. RSTV is
   inscribed in a circle, it is also known as a cyclic _____.

4. While quadrilateral RSTV is said to be inscribed in ⊙ Q,
   this circle is said to be _____ about RSTV.

5. Complete: If a quadrilateral is cyclic, then its opposite angle are _____.

6. Complete: A polygon is circumscribed about a circle if its sides are _____ to the circle.

More Angles in the Circle

7. Complete: The measure of an angle formed by two chords that intersect within a circle is one-half
   the sum of measures of the arcs intercepted by the angle and its _____ angle.

8. Fill the blanks to complete the proof of the theorem of Exercise 7.

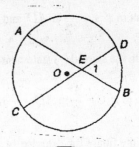

   Given: In ⊙ O, chords $\overline{AB}$ and $\overline{CD}$ intersect at E

   Prove: $m\angle 1 = \frac{1}{2}(m\overset{\frown}{AC} + m\overset{\frown}{DB})$

   Proof: In ⊙ O, chords $\overline{AB}$ and $\overline{CD}$ intersect at E. Draw
   chord ____. Because ∠ 1 is an exterior angle of Δ CBE,
   it follows that $m\angle 1 =$ _____ + _____. (*)

   But $m\angle 2 = \frac{1}{2}m\overset{\frown}{DB}$ and $m\angle 3 =$ _____.

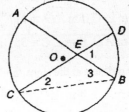

   Substitution into the starred (*) statement leads to

   $m\angle 1 = \frac{1}{2}m\overset{\frown}{AC} + \frac{1}{2}m\overset{\frown}{DB}$, so $m\angle 1 =$ _____.

9. In the figure for Exercise 8, $m\overset{\frown}{AC} = 68^0$ and $m\overset{\frown}{DB} = 44^0$.
   Find: (a) m∠ 1 _____ (b) m∠ CEB _____.

10. Complete: The radius drawn to a tangent at its point of contact is _____ to the tangent.

Angles Formed by A Tangent and Chord

11. Complete: The measure of an angle formed by a tangent and a chord drawn to the point of tangency is equal to _____ the measure of the intercepted arc.

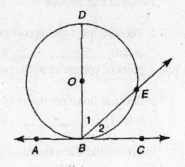

12. In the drawing, $\overrightarrow{AC}$ is tangent to $\odot O$ and $\overline{DB}$ is a diameter.

If m $\overset{\frown}{EB}$ = 82^0, find: (a) m$\angle 2$ _____ (b) m$\angle 1$ _____.

(c) m$\angle ABD$ _____ (d) m$\overset{\frown}{DE}$ _____.

Angles with the Vertex in the Exterior of the Circle

13. Complete: The measure of an angle formed when two secants intersect at a point outside the circle is one-half the _____ of the measures of the two intercepted arcs.

14. Fill the blanks to complete the proof of the theorem of Exercise 13.

Given: In the circle, secants $\overline{AC}$ and $\overline{CD}$ intersect at C

Prove: m$\angle C = \dfrac{1}{2}$ (m $\overset{\frown}{AD}$ - m $\overset{\frown}{BE}$)

Proof: In the circle, secants $\overline{AC}$ and $\overline{CD}$ intersect at C. Draw chord ____ to form $\triangle$ BCD. Because $\angle 1$ is an exterior angle of $\triangle CBD$, it follows that m$\angle 1 =$ _____ + _____. (*)

But m$\angle 1 = \dfrac{1}{2}$ m $\overset{\frown}{AD}$ and m$\angle D =$ _____. Substitution

into the starred (*) statement leads to $\dfrac{1}{2}$ m $\overset{\frown}{AD}$ = m$\angle C + \dfrac{1}{2}$ m $\overset{\frown}{BE}$.

So m$\angle C = \dfrac{1}{2}$ m $\overset{\frown}{AD}$ - $\dfrac{1}{2}$ m $\overset{\frown}{BE}$ and m$\angle C =$ _____.

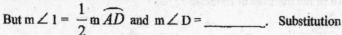

15. In the figure for Exercise 14, find:

(a) m$\angle C$ if m $\overset{\frown}{AD}$ = 117^0 and m $\overset{\frown}{BE}$ = 39^0 _____.

(b) m $\overset{\frown}{AD}$ if m$\angle C$ = 35^0 and m $\overset{\frown}{BE}$ = 37^0. _____.

16. In the figure at the right, find:

(a) m$\angle L$ if m $\overset{\frown}{JH}$ = 125^0 and m $\overset{\frown}{JK}$ = 47^0 _____.

(b) m $\overset{\frown}{JK}$ if m$\angle L$ = 38^0 and m $\overset{\frown}{JH}$ = 3(m $\overset{\frown}{JK}$). _____.

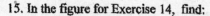

17. In the figure, find m $\angle 1$ if m $\overset{\frown}{ACB}$ = 235^0. _____.

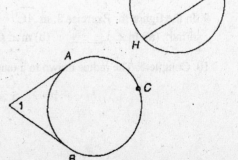

60

Three Related Theorems

1. Complete: If a line through the center of a circle is perpendicular to a chord, then it bisects the _____ and its _____.

2. Complete: If a line through the center of a circle bisects a chord (other than a diameter), then the line is _____ to the chord.

3. Complete: The perpendicular-_____ of a chord contains the center of the circle.

4. The radius of ⊙ O has the length 10.

   Also, $\overline{OE} \perp \overline{CD}$ at B and OB = 6. Find:
   (a) CB _____ (b) CD _____.

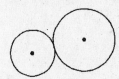

Tangent Circles and Tangents to Circles

5. The circles shown at the right are _____ tangent circles.

6. How many common tangents do each pair of circles have?
   (a) ___   (b) ___   (c) ___.

Tangent Segments to Circles

7. Complete: The tangent line segments drawn to a circle from an external point are _____.

8. Fill the blanks to complete the proof of the theorem in Exercise 7.

   Given: $\overline{AB}$ and $\overline{AC}$ are tangents to ⊙ O from point A

   Prove: $\overline{AB} \cong \overline{AC}$

   Proof: Draw chord $\overline{BC}$ to form △ ABC. Now m ∠ B = $\frac{1}{2}$ m $\overset{\frown}{BC}$

   and m ∠ C = _____. Then ∠ B ≅ ∠ C and it follows that _____ since these sides lie opposite congruent angles of △ ABC.

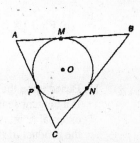

9. With ⊙ O inscribed in △ ABC, suppose that AP = 5.2, CN ≈ 4.3, and MB = 7.1.
   Find: (a) AM _____ (b) AC _____ (c) BC _____ (d) AB _____.

10. In the figure for Exercise 9, suppose that AB = 14, BC = 16, and AC = 12.
    Let AM ≈ x, CP = y, and BN ≈ z; based upon the lengths of the sides of △ ABC, write three equations in terms of variables x, y, and z.

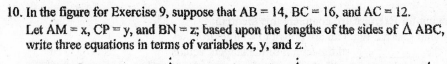

    [Hint: Each equation contains 2 of the variables.]

    Exercises 9 & 10

61

11. Complete: If two chords intersect within a circle, the product of the lengths of segments (parts) of one chord is equal to the _____ of the lengths of the segments of the other chord.

12. Fill the blanks to complete the proof of the theorem in Exercise 11.

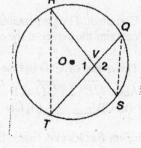

Given: ⊙ O with chords $\overline{RS}$ and $\overline{TQ}$ intersecting at V.
Prove: RV · VS = TV · VQ

Proof: ⊙ O with chords $\overline{RS}$ and $\overline{TQ}$ intersecting at V. Now draw chords $\overline{RT}$ and $\overline{QS}$ to form △ RTV and △ _____. In the triangles, ∠1 ≅ ∠2 because these are a pair of _____ angles. Also, ∠R ≅ ∠___ because these inscribed angles intercept the same arc. Then △ RTV ~ △ QSV by the reason _____. Using CSSTP, it follows that $\dfrac{RV}{VQ} = \dfrac{TV}{VS}$. By the Means-Extremes Property for Proportions, we conclude that _____.

13. Use the figure for Exercise 12.
    (a) Find VS if RV = 8, TV = 7, and VQ = 6. _____.

    (b) Find RV if TV = 6, VQ = 4, and RS = 11. _____.

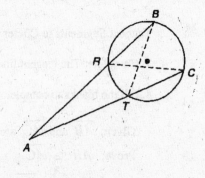

14. In the drawing at right, we can show that △ ABT ~ △ ACR; As a consequence, AB · RA = AC · TA. Find AB if AR = 9, AT = 8, and TC = 10. _____.

15. Based upon the figure for Exercise 14, complete this theorem: If two secant segments are drawn to a circle from an external point, then the products of the length of each secant with the length of its _____ segment are equal.

16. In the drawing at right, we can show that △ TVW ~ △ TXV and, in turn, that $(TV)^2 = TW \cdot TX$.
    (a) Find TV if TW = 18 and TX = 8 _____.

    (b) Find TX if TV = $2\sqrt{15}$, TX = x, and TW = x + 4. _____.

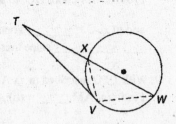

17. Based upon the drawing for Exercise 16, complete this theorem: If a tangent segment and a secant segment are drawn to a circle from an external point, then the _____ of the length of the tangent equals the product of the length of the secant with the length of its external segment.

Construction of tangents to Circles

1. The line that is perpendicular to the radius of a circle at its
   endpoint on the circle is a(n) _____ to the circle.

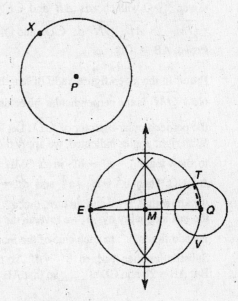

2. Use information from the theorem of Exercise 1 to
   construct the tangent to ⊙ P at point X.

3. The construction of the tangent(s) to ⊙ Q
   from external point E can be justified by the
   fact that ∠ETQ is a right angle. Then $\overline{ET} \perp \overline{TQ}$.

   According to the theorem in Exercise 1, $\overline{ET}$
   is a(n) _____ to ⊙ Q.

Inequalities in a Circle

4. Complete: In a circle (or congruent circles) containing two unequal central angles,
   the larger central angle corresponds to the _____ intercepted arc.

5. Fill the blanks for the proof of the theorem of Exercise 4.

   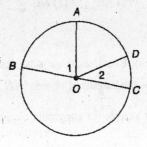

   Given: ⊙ O with central ∠s 1 and 2; m∠1 > m∠2
   Prove: m $\overarc{AB}$ > m $\overarc{CD}$

   Proof: In ⊙ O, m∠1 > m∠2. But m∠1 = m $\overarc{AB}$
   and m∠2 = _____. By substitution, _____.

6. The converse of the theorem in Exercise 4 is also true. Using the figure
   in Exercise 5 and the following information, draw conclusions.

   (a) m $\overarc{AB}$ > m $\overarc{CD}$ _____ (b) m∠1 = 73° and m∠2 = 38° _____

7. Complete: In a circle (or congruent circles) containing two unequal chords,
   the _____ chord is farther from the center of the circle.

8. Complete: In a circle (or congruent circles) containing two unequal chords,
   the chord that is nearer the center of the circle has the _____ length.

9. In the drawing, RP = 8, PS = 7, and TP = 4. Using the information
   found in Exercises 7 and 8, which chord is:
   (a) longest? _____ (b) shortest? _____.

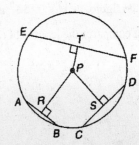

63

10. Fill in the blanks for the proof of the theorem in Exercise 8.

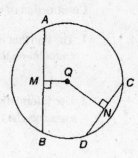

Given: $\odot$ Q with chords $\overline{AB}$ and $\overline{CD}$;
$\overline{QM} \perp \overline{AB}$, $\overline{QN} \perp \overline{CD}$, and QM < QN
Prove: AB > CD

Proof: In the given figure, radii of length r are drawn as shown.
Now $\overline{QM}$ is the perpendicular-bisector of _____ and _____ is

the perpendicular-bisector of $\overline{CD}$. Let MB = b and NC = d.
With right angles indicated, we apply the Pythagorean Theorem
to show that $r^2 = a^2 + b^2$ in $\triangle$ QMB. Also, $r^2 =$ _____ in
$\triangle$ QNC. Then $b^2 = r^2 - a^2$ and $d^2 =$ _____. With QM < QN,
we know that a < c and, in turn, that $a^2 < c^2$. Multiplying each side of
the last inequality by -1, we reverse the inequality to obtain _____.
By addition of r^2 to each side of the inequality, $r^2 - a^2 > r^2 - c^2$.
Substitution then leads to $b^2 > d^2$, so that b > ____; then 2b > 2d.
But AB = 2b and CD = ____, so that AB > CD.

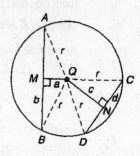

11. In the figure for Exercise 10, suppose that the length of radius for $\odot$ Q is 5.
If QM = 3 and QN = 4, find:
(a) AM ____ (b) AB ____ (c) NC ____ (d) CD ____

12. In the figure for Exercise 10, suppose that QN = 5 and QM = 4. State an inequality
that compares the lengths of chords $\overline{AB}$ and $\overline{CD}$. _____.

13. Complete: In a circle (or congruent circles) containing two unequal chords,
the _____ chord corresponds to the greater minor arc.

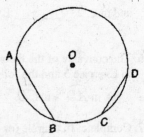

14. In the figure, AB = 6.8 and CD = 4.9. State an inequality that
compares m $\overarc{AB}$ and m $\overarc{CD}$. _____.

15. Complete: In a circle (or congruent circles) containing two unequal minor arcs, the
greater minor arc corresponds to the _____ chord related to these arcs.

16. In the figure, m $\overarc{MN}$ = 58° and m $\overarc{PQ}$ = 29°. Write an inequality
that compares the lengths of chords $\overline{MN}$ and $\overline{PQ}$. _____.

Locus of Points

1. Definition: A _____ of points is the set of all points and only those points that satisfy a given condition or set of conditions.

2. Draw and describe "the locus of points in a plane that are at a distance of 1 inch from a given point P."
   The locus is _____
   _____.

P.

3. Suppose that problem 2 had asked for "the locus of points in *space* that are at a distance of 1 inch from a given point P." What type of figure would then be determined? _____

4. Draw and describe "the locus of points in a plane that are at a distance of 1 centimeter from a given line ℓ ."
   The locus is _____
   _____.

An Important Theorem

5. Complete: The locus of points in a plane that are equidistant from the sides of an angle is _____.

6. Fill the blanks for the 2-part proof of the theorem in Exercise 5.
   (i.) If a point is on the bisector of an angle, then it is equidistant from the sides of the angle.

   Given: $\overline{BD}$ bisects $\angle ABC$; $\overline{DE} \perp \overline{BA}$ and $\overline{DF} \perp \overline{BC}$
   Prove: $\overline{DE} \cong \overline{DF}$

   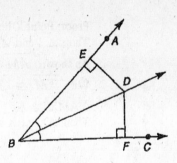

   Proof: $\overline{BD}$ bisects $\angle ABC$ so that $\angle ABD \cong \angle$ _____.
   $\overline{DE} \perp \overline{BA}$ and $\overline{DF} \perp \overline{BC}$, so $\angle DEB$ and $\angle DFB$ are
   right angles. Then $\angle DEB \cong \angle$ _____. By the reason "Identity,"
   we have _____. Then $\triangle DEB \cong \triangle$ _____ by AAS.
   Finally, _____ by CPCTC.

   (ii.) If a point is equidistant from the sides of an angle, then it lies on the bisector of the angle.

   Given: $\angle ABC$ with $\overline{DE} \perp \overline{BA}$ and $\overline{DF} \perp \overline{BC}$;
          also, $\overline{DE} \cong \overline{DF}$
   Prove: $\overline{BD}$ bisects $\angle ABC$

   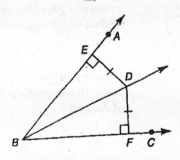

   Proof: $\overline{DE} \perp \overline{BA}$ and $\overline{DF} \perp \overline{BC}$, so $\triangle$ _____ and
   $\triangle$ _____ are right triangles. By hypothesis, ____ $\cong$ ____.
   By "Identity," _____, Then $\triangle DEB \cong \triangle DFB$ by HL.
   Then $\angle$ ____ $\cong \angle$ ____ by CPCTC so that
   _____ by the definition of angle-bisector.

Another Important Theorem

7. Complete: The locus of points in a plane that are equidistant from the endpoints of a line segment is the _____

8. Fill the blanks for the 2-part proof of the theorem in Exercise 7.
   (i) If a point is equidistant from the endpoints of a line segment, then it lies on the perpendicular-bisector of the line segment.

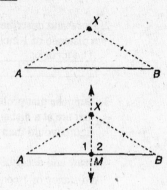

   Given: $\overline{AB}$ and point X so that AX = BX
   Prove: X lies on the perpendicular-bisector of $\overline{AB}$

   Proof: Let M name the midpoint of $\overline{AB}$, so that $\overline{AM} \cong$ ____.
   With AX = BX, we also know that $\overline{AX} \cong$ _____. By "Identity,"
   _____. Then $\triangle AMX \cong \triangle$ ____ by SSS. Using
   CPCTC, we know that $\angle 1 \cong \angle$ ___; then ___ $\perp$ ___.
   By definition, $\overline{MX}$ is the perpendicular-bisector of $\overline{AB}$ so it
   follows that _____.

   (ii) If a point is on the perpendicular-bisector of a line segment, then the point is equidistant from the endpoints of the line segment.

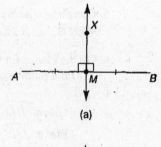

(a)

   Given: Point X lies on the perpendicular-bisector of $\overline{AB}$
   Prove: AX = BX (Point X is equidistant from A and B)

   Proof: Point X lies on the perpendicular-bisector of $\overline{AB}$
   Then $\angle 1$ and $\angle 2$ are congruent _____ angles.
   As shown, $\overline{AM} \cong$ __ because $\overleftrightarrow{XM}$ bisects $\overline{AB}$.
   With $\overline{XM} \cong \overline{XM}$ by "Identity," it follows that
   $\triangle AMX \cong \triangle$ _____ by SAS. By CPCTC, $\overline{XA} \cong \overline{XB}$.
   Then AX = BX so that X is equidistant from A and B.

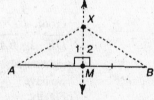

Constructions that use the Locus Concept

9. The figure at the right represents "the locus of the vertex of
   a right triangle with hypotenuse $\overline{AB}$." Why must each angle
   marked with the square at its vertex be a right angle?

   _____.

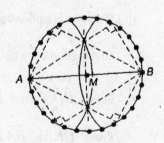

10. Given its diagonals $\overline{AC}$ and $\overline{BD}$, construct rhombus ABCD.
    Note: The diagonals of a rhombus are perpendicular-bisectors
    of each other.

Concurrent Lines

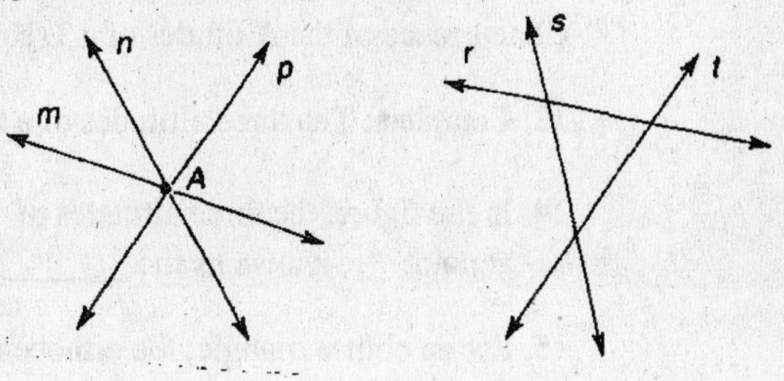

1. Complete: A number of lines are _____ if they have exactly one point in common.

2. In the figure shown, which group of lines (m, n, and p *or* r, s, and t) are concurrent? _____.

Concurrence of the Angle-Bisectors of a Triangle

3. Complete: The three angle-bisectors of a triangle are _____.

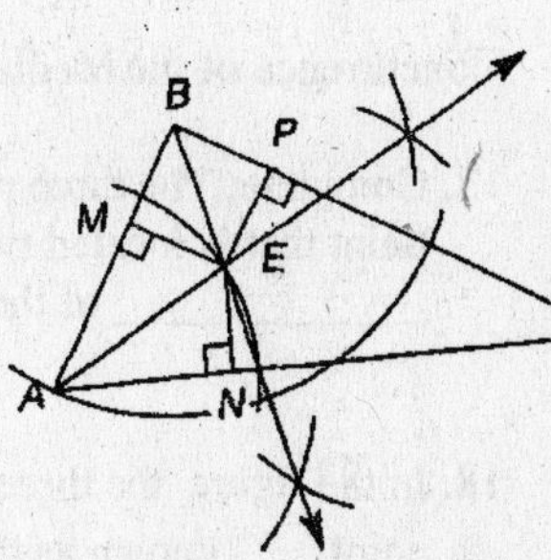

4. In the figure, the three angle-bisectors shown in △ ABC are concurrent at point ___, known as the _____ of △ ABC.

5. Based upon the locus theorem, "The locus of points equidistant from the sides of an angle is the bisector of the angle," we know that point E is _____ from the 3 sides of △ ABC.

6. Suggested by the congruence of $\overline{EN}$, $\overline{EP}$, and $\overline{EM}$ in △ ABC, it is always possible to _____ a circle within a triangle.

7. In order to locate the incenter of a triangle, how many of the angle-bisectors of the triangle must be constructed? _____.

Concurrence of the Perpendicular-Bisectors of Sides of a Triangle

8. Complete: The three perpendicular-bisectors of the sides of a triangle are _____.

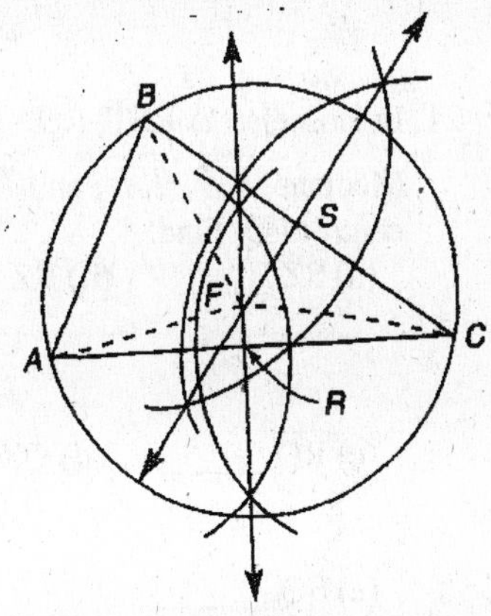

9. In the figure, the three perpendicular-bisectors of the sides of △ ABC are concurrent at point ___, known as the _____ of △ ABC.

10. Based upon the locus theorem, "The locus of points equidistant from the endpoints of a line segment is the perpendicular-bisector of the line segment," we know that point F is equidistant from the three _____ of △ ABC.

11. Suggested by the congruence of $\overline{AF}$, $\overline{BF}$, and $\overline{CF}$ in △ ABC, it is always possible to _____ a circle about a triangle.

12. In order to locate the circumcenter of a triangle, how many of the perpendicular-bisectors of sides of the triangle must be constructed? _____

67

Concurrence of the Altitudes of a Triangle

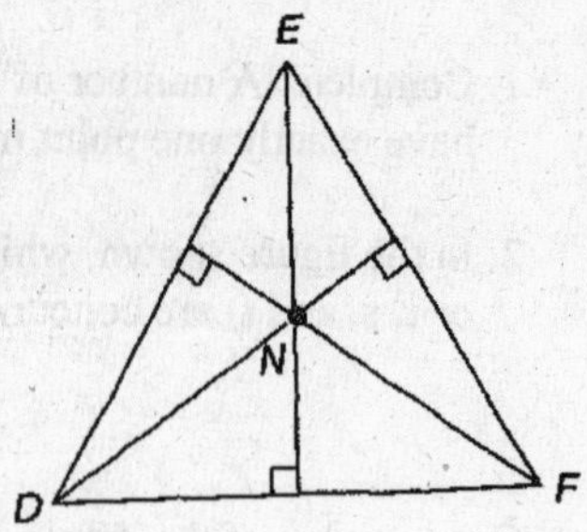

13. Complete: The three altitudes of a triangle are _____.

14. In the figure, the three altitudes of △ DEF are concurrent
at point ___, known as the _____ of △ DEF.

15. For an obtuse triangle, the orthocenter lies in the _____ of
the triangle.

16. In order to locate the orthocenter of a triangle, how many of the
altitudes of the triangle must be constructed? _____.

Concurrence of the Medians of a Triangle

17. Complete: The three medians of a triangle are _____ at a
point that is located two-thirds of the distance from a vertex to the
_____ of the opposite side.

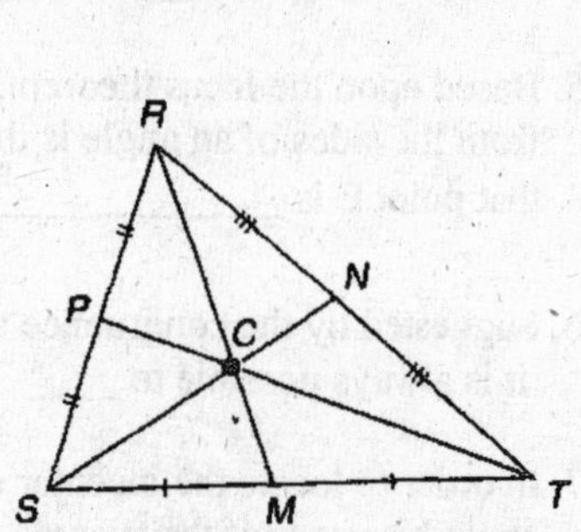

18. In the figure, the three medians of △ RST are concurrent at
point ___, known as the _____ of △ RST.

For Exercises 19 and 20, use the figure from Exercise 18.

19. In △ RST, the length of the median $\overline{RM}$ is 12 inches. Find:
(a) RC _____ (b) CM _____.

20. In △ RST, the length of $\overline{TC}$ is 10 centimeters. Find:
(a) TP _____ (b) CP _____.

21. In isosceles △ RST, RS = RT = 15, and ST = 18.
Medians $\overline{RZ}$, $\overline{TX}$, and $\overline{SY}$ are concurrent at
centroid Q. Find: 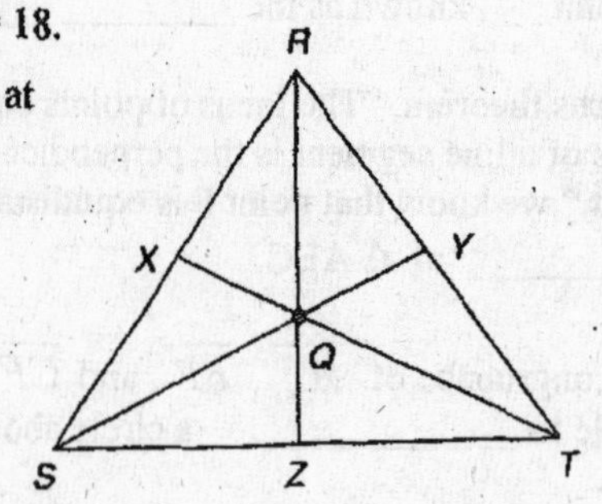
(a) SZ _____ (b) RZ _____.

(c) RQ _____ (d) QZ _____.

(e) SQ _____.

Polygons with Inscribed Circles

1. Because the angle-bisectors of a triangle are concurrent at a point (incenter of triangle) that is equidistant from the sides of the triangle, it is always possible to _____ a circle within the triangle.

2. Does the following quadrilateral have concurrent angle-bisectors and thus an inscribed circle?
   (a) any rectangle _____
   (b) any rhombus _____.
   (c) any square _____
   (d) any isosceles trapezoid _____.

3. Which of these *regular* polygons have concurrent angle-bisectors and, in turn, an inscribed circle?
   _____

Polygons with Circumscribed Circles

4. Because the perpendicular-bisectors of the sides of a triangle are concurrent at a point (circumcenter of triangle) that is equidistant from the vertices of the triangle, it is always possible to _____ a circle about the triangle.

5. Does the following quadrilateral have concurrent perpendicular-bisectors of its sides and thus a circumscribed circle?
   (a) any rectangle _____
   (b) any rhombus _____.
   (c) any square _____
   (d) any isosceles trapezoid _____.

6. Which of these *regular* polygons have concurrent perpendicular-bisectors of sides and, in turn, a circumscribed circle? _____

Angles in Regular Polygons

7. Use $I = \dfrac{(n-2) \cdot 180^0}{n}$ to find the measure of each interior angle of a regular pentagon. _____.

8. Use $E = \dfrac{360^0}{n}$ to find the measure of each exterior angle of a regular octagon. _____.

More about Regular Polygons

9. Complete: A circle can be inscribed within or circumscribed about any _____ polygon.

10. For a regular polygon, how are the center of its inscribed circle and the center of its circumscribed circle related? _____.

11. Complete (definition): The _____ of a regular polygon is the common center for the inscribed circle and the circumscribed circle of the polygon.

12. Complete (definition): A(n) _____ of a regular polygon is any line segment joining the center of the regular polygon to one of its vertices.

13. Complete: Any radius of a regular polygon _____ the angle to which it is drawn.

14. How many radii does a regular decagon (10 sides) have? _____.

15. Find the length of radius for a square that has sides of length 6.8 cm each. _____.

16. Complete (definition): A(n) _____ of a regular polygon is any line segment drawn from the center of the regular polygon so that it is perpendicular to a side.

17. Complete: Any apothem of a regular polygon _____ the side to which it is drawn.

18. Find the length of the apothem of a regular hexagon whose sides are each 6 feet in length. _____.

19. Find the length of each side of an equilateral triangle for which each apothem has a length of 2 cm. _____.

20. Find the perimeter of a regular hexagon whose length of radius is $4\sqrt{5}$ inches. _____.

Initial Area Concepts

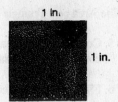

1 in.

1 in.

1. While linear units such as inches or centimeters are used to measure length or distance, the unit of area shown measures one _____ inch. For convenience, we represent 1 square _____ by the symbol in^2.

2. Complete (Area Postulate): Corresponding to every bounded region is a unique positive number A known as the _____ of the region.

3. Complete: If two closed plane regions are congruent, then their areas are _____.

4. Complete (Area-Addition Postulate): Given that A_R and A_S are the areas of nonoverlapping regions R and S, $A_{R \cup S} =$ _____.

5. In the figure, $A_R = 13.4 \ cm^2$ and $A_S = 6.9 \ cm^2$. Find $A_{R \cup S}$. _____

The Area of a Rectangle

6. By counting the number of square centimeters, find the area of rectangle MNPQ. $A_{MNPQ} =$ _____.

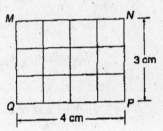

3 cm

4 cm

7. In rectangle MNPQ, the base (or length) measures b = 4 cm while the altitude (or width) measures h = 3 cm. Use the formula A = bh (or A = ℓ w) to find the area of rectangle MNPQ. $A_{MNPQ} =$ _____.

8. Find the area (in square inches) of a rectangle that has length ℓ = 1 foot and width w = 5 inches. _____.

9. How many square inches are in one square foot? _____.

10. If each side of a square has length s, then the formula for its area is A = _____.

The Area of a Parallelogram

b

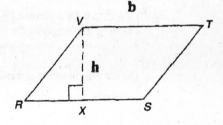

h

11. In the figure provided, it is shown that the area of the parallelogram with base length b and corresponding length of altitude h can be found by the formula A = _____.

12. Find the area of ▱ RSTV if RS = 7 cm, RX = 3 cm, and VR = 5 cm. A = _____.

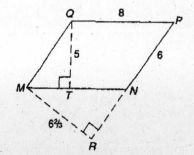

8

5

6

6⅔

13. In ▱ MNPQ, what is the length of the altitude that corresponds to:

 (a) base $\overline{MN}$? ____ (b) base $\overline{NP}$? ____.

14. Find the area of ▱ MNPQ. _____

71

The Area of a Triangle

15. Complete: The area A of a triangle whose base has length b and whose corresponding altitude has length h is given by A = _____.

16. Fill the blanks in the proof for the theorem in Exercise 15.

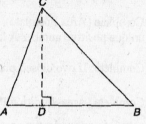

   Given: $\triangle ABC$ with $\overline{CD} \perp \overline{AB}$;
   Also, AB = b and CD = h

   Prove: $A_{\triangle ABC} = \dfrac{1}{2} bh$

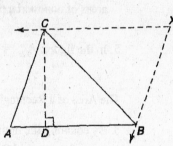

   Proof: In $\triangle ABC$, we construct $\overrightarrow{BX} \parallel \overline{AC}$ and $\overrightarrow{CX} \parallel$ _____
   as shown. Then quad. ACXB is a _____ and
   it follows that $A_{\square ACXB} = b \cdot h$.

   With $\overline{CB}$ as a diagonal of $\square$ ABXC, $\triangle$ ABC $\cong \triangle$ _____.

   Because $A_{\triangle ABC} = A_{\triangle XCB}$, we have $A_{\square ABXC} = A_{\triangle ABC} + A_{\triangle ABC}$

   or $A_{\square ABXC} = 2 \cdot A_{\triangle ABC}$. Multiplying by $\dfrac{1}{2}$, $A_{\triangle ABC} = \dfrac{1}{2} A_{\square ABXC}$.

   Equivalently, $A_{\triangle ABC} =$ _____.

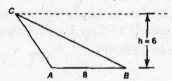

17. Find the area of $\triangle$ ABC, which is drawn
    at the right. _____.

18. Complete: The area of a right triangle whose legs have
    lengths a and b is given by the formula A = _____.

19. Find the area of $\triangle$ PMN if PM = 8 cm and PN = 6 cm. _____.

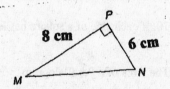

20. In the figure for Exercise 19, find:
    (a) MN, the length of the hypotenuse. _____.

    (b) the length of the altitude (not shown) from
        vertex P to hypotenuse $\overline{MN}$. _____.

Perimeter of Polygons

1. Complete: The _____ of a polygon is the sum of the lengths of its sides.

2. Complete: When the area of a polygon is measured in square meters, we would probably measure the perimeter of that polygon in _____.

3. Complete these formulas for the perimeter of:
   (a) an equilateral triangle with a side of length s. $P =$ _____.
   (b) a rectangle whose base has length b and whose altitude has length h. $P =$ _____.

4. Find the perimeter of:
   (a) an isosceles triangle with a leg of length 5 cm and a base of length 6 cm. _____

   (b) rhombus ABCD with diagonals of length $d_1 = 6$ ft and $d_2 = 8$ ft. _____.

   (c) square RSTV whose area measures 64 in^2 _____.

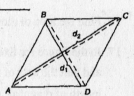

Heron's Formula

5. Complete: The semiperimeter s of a triangle with sides of lengths a, b, and c is $s =$ _____.

6. Complete (Heron's Formula): For the triangle with lengths of sides a, b, and c and semiperimeter s, the _____ is given by the formula $A = \sqrt{s(s-a)(s-b)(s-c)}$.

7. Use Heron's Formula (See Exercise 6) to find the area of the triangle that has sides of lengths a = 3 cm, b = 4 cm, and c = 5 cm. _____.

8. Use Heron's Formula (See Exercise 6) to find the area of the triangle that has sides of lengths a = 5 in, b = 6 in, and c = 7 in. _____.

The Area of a Trapezoid

9. Complete: By drawing an auxiliary line (a diagonal), the area of a trapezoid can be shown to be equal to the sum of the areas of two _____.

10. Complete: The area of a trapezoid with bases of lengths b_1 and b_2 and whose altitude has length h is given by $A =$ _____.

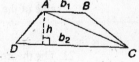

11. Find the area of trapezoid ABCD if $b_1 = 5$ in, $b_2 = 13$ in, and h = 4 in. _____. Exercises 10 & 11

12. In trapezoid RSTV (shown), RS = 6 cm and VT = 14 cm. If the area of trapezoid RSTV is 70 cm^2, find the length of altitude $\overline{RW}$. _____

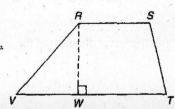

73

Quadrilaterals with Perpendicular Diagonals

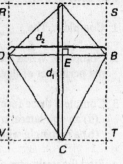

13. In the figure, quadrilateral ABCD has perpendicular diagonals of lengths d_1 and d_2. With auxiliary lines drawn parallel to the diagonals as shown, quad. RSTV is a(n) _____.

14. Complete: The area of any quadrilateral with perpendicular diagonals of lengths d_1 and d_2 is given by the formula $A =$ _____.

15. Two specific types of quadrilaterals that have perpendicular diagonals are the _____ and the _____.

16. Find the area of rhombus MNPQ if QN = 10 ft and PM = 8 ft. _____

17. In the figure for Exercise 16, suppose that QN = 24 inches and that the area of rhombus MNPQ is 240 in^2. Find the length of diagonal $\overline{PM}$. _____

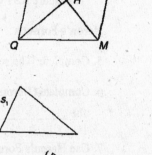

Areas of Similar Polygons

18. For the triangles shown, the lengths of the sides of the larger triangle are twice the lengths of the corresponding sides of the smaller triangle; that is, $\frac{s_2}{s_1} = \frac{2}{1}$. However, the ratio of their areas is $\frac{A_2}{A_1} =$ _____.

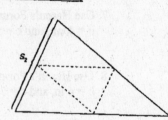

19. Complete: The ratio of the areas of two similar triangles equals the square of the ratio of the lengths of two _____ sides. If the lengths of two corresponding sides are s_1 and s_2, then the ratio of their areas is $\frac{A_2}{A_1} =$ _____.

20. Find the ratio of the areas of two similar triangles if $a_1 = 6$ inches in the smaller triangle while $a_2 = 9$ inches is the length of the corresponding side of the larger triangle. $\frac{A_2}{A_1} =$ _____.

21. The theorem stated in Exercise 19 is valid for any two similar polygons. For the squares shown, the lengths of the sides of the smaller square are one-third the length of the sides of the larger square; that is, $\frac{s_2}{s_1} = \frac{1}{3}$.

However, the ratio of the areas of the squares is $\frac{A_2}{A_1} =$ _____.

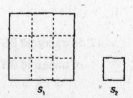

The Area of a Square

1. Write the formula for the area A of a square in terms
   of the length s of each of its sides. _____.

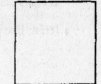

2. Find the area of a square that has a perimeter of 36 inches. _____.

3. Find the area of a square that has an apothem of length 5 cm. _____.

4. Find the length of each side of a square that has an area of 81 ft^2. _____.

The Area of an Equilateral Triangle

5. Write the formula for the area A of an equilateral triangle in terms
   of the length s of each of its sides. _____.

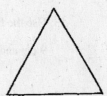

6. Find the area of an equilateral triangle that has a perimeter of 30 inches. _____.

7. Find the area of an equilateral triangle that has a radius of length $4\sqrt{3}$ cm. _____.

8. Find the length of each side of an equilateral triangle that has an area of $9\sqrt{3}$ ft^2. _____.

The Area of a Regular Polygon

9. When the five radii of a regular pentagon are drawn, the pentagon is separated
   into five congruent triangles. With each side of length s and each apothem of
   length a, the area of the regular pentagon is the sum of areas of the triangles,

   $$A = \frac{1}{2}as + \frac{1}{2}as + \frac{1}{2}as + \frac{1}{2}as + \frac{1}{2}as \quad \text{or} \quad A = \frac{1}{2}a(s+s+s+s+s)$$

   or A = _____, where P is the perimeter of the regular pentagon.

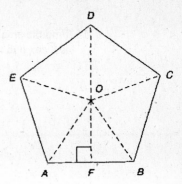

10. For a square whose sides are of length s, show that the formula $A = \frac{1}{2}aP$

    is equivalent to the formula $A = s^2$. Solution: $A = \frac{1}{2}aP = $ _____.

    [Hint: For a square, $a = \frac{1}{2}s$ and $P = 4s$.]

75

11. For an equilateral triangle whose sides are of length s, show that the formula $A = \dfrac{1}{2} aP$

    is equivalent to the formula $A = \dfrac{s^2}{4} \sqrt{3}$. Solution: $A = \dfrac{1}{2} aP =$ _____.

    [Hint: Use the same approach as described in the "Hint" of Exercise 10.]

12. Use the formula $A = \dfrac{1}{2} aP$ to find the area of a regular hexagon in which each side has the length

    10 centimeters. _____.

13. Use the formula $A = \dfrac{1}{2} aP$ to find the area of a regular pentagon in which each side has the length

    9 cm and for which the apothem measures 6.2 cm. _____.

14. Find the area of a regular octagon in which each side measures 6.7 feet and for which the length
    of the apothem is 8.2 feet. _____.

15. Find the area of a regular polygon in which each central angle measures 30°, the length of each side
    is 8 cm, and the length of the apothem is 15 cm. _____.

The Value of π

1. Complete: The ratio of the _____ C of a circle to the length of its _____ d

   is a unique positive number; for the ratio $\dfrac{C}{d}$, we use the symbol _____.

2. Approximations of the value of ____ include the decimal 3.1416 and the common fraction _____.

The Circumference of a Circle

3. Based upon the fact $\dfrac{C}{d} = \pi$, we derive these formulas for the circumference of a circle:

   $$C = \underline{\quad} \text{ and in turn, } C = 2\pi r$$

4. For the circle shown, the length of radius is r = 7 inches. Find:
   (a) the exact circumference _____.
      [Note: Leave π in the answer.]

   (b) the approximate circumference _____.

      [Note: Let $\pi \approx \dfrac{22}{7}$.]

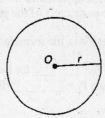

5. The circumference of a circle (not shown) is approximately 44 cm.
   To the nearest whole number, find the length of the:
   (a) radius of the circle _____ (b) the diameter of the circle _____.

The Length of an Arc

6. m $\overset{\frown}{AB}$ represents the degree measure of an arc, while $\ell\ \overset{\frown}{AB}$ represents the _____ of that arc.

7. Suppose that m $\overset{\frown}{AB}$ = 90° in the circle shown. What common fraction
   names the part of the circumference that is the length of arc AB? _____.

8. In $\odot$ O, m $\overset{\frown}{AB}$ = 90° and the length of radius is OA = 8 cm. Find:

   (a) the exact length represented by $\ell\ \overset{\frown}{AB}$. _____.

   (b) the approximate length represented by $\ell\ \overset{\frown}{AB}$,
      correct to nearest tenth of a centimeter. _____.

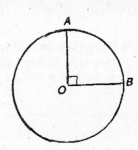

Limits

9. For some measurements, there are both a largest and a smallest possible value. Such numbers are known
   as the upper limit and the _____ limit, respectively.

10. In a circle of radius length 5 cm, the upper limit for
    the length of a chord is _____ cm while the lower
    limit for the length of a chord is _____ cm.

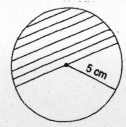

5 cm

77

11. In ⊙ O with diameter $\overline{DC}$, one side of the inscribed angle with
vertex A is chord $\overline{DA}$. If point B is any point between points A
and C, what is the lower limit for the measure m ∠ DAB? _____ .

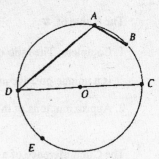

The Area of a Circle

12. As the number of sides of regular polygons inscribed in a circle increases,
the areas of the regular polygons also increase and approach the area of
the _____ . As the number of sides increases, the length of
the apothem approaches the length of the _____ of the circle
while the perimeters of the polygons approach the _____ of

the circle. Thus, the formula for the area of the regular polygon $A = \dfrac{1}{2} aP$

becomes A = _____ for the circle. Substitution of $2\pi r$ for C leads to
the formula A = _____ for the area of the circle.

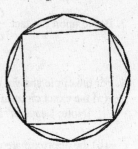

13. Find the exact area for a circle whose radius measures 4 cm. _____ .
[Note: Leave π in the answer.]

14. Where $\pi \approx 3.14$, find the approximate area for a circle whose radius is r = 6.7 inches.
Give answer correct to nearest tenth of a square inch. _____ .

15. Where $\pi \approx \dfrac{22}{7}$, find the approximate area of the circle whose diameter has the

length d = 7 inches. Give the answer as a mixed number. _____ .

16. Using the calculator value of π , find the area of the semicircular region
for which the length of radius is r $\approx$ 3.47 meters. Give the answer correct
to the nearest tenth of a square meter. _____ .

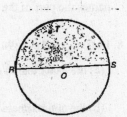

17. Using the calculator value of π , find the area of the shaded region
bounded by concentric circles. For this ring (annulus), r = 2.3 cm
and R = 4.2 cm. Give the answer correct to the nearest tenth of a
square centimeter. _____ .

Sector of a Circle

1. Complete (definition): A _____ of a circle is a region that is bounded by two radii and an arc intercepted by those radii.

2. Considering the measure of the central angle (and intercepted arc), what common fraction indicates the part of the area of circle that is the area of the sector?

   (a) _____   (b) _____

3. Complete (theorem): In a circle of radius r, the area of a sector whose arc has measure m is given by $A =$ _____ .

4. Find the approximate area of a sector if the radius of the circle is 8.6 inches and the central angle measures 110^0. Give the answer correct to the nearest tenth of a square inch. _____ .

5. For the sector described and shown in Exercise 4, the plan (formula) for calculating its perimeter is $P = 2r +$ ____ .

6. Find the exact perimeter of a sector (not shown) for a circle of radius 6 feet if the measure of the arc of the sector is 120^0. _____ .

Segment of a Circle

7. Complete (definition): A _____ of a circle is a region bounded by a chord and the arc with the same endpoints as the chord.

8. The plan for determining the area of a segment is given by

   $A_{segment} =$ _____ - _____ .

9. Find the exact area of the segment in a circle of radius r = 12 in if the related central angle (and arc) measures 90^0. _____ .

10. For the segment described (and shown) in Exercise 9, find the exact perimeter. _____ .

11. Find the exact area of the larger segment of Exercise 9. For that circle, r = 12 in and the segment has a related major arc that measures 270^0. _____ .

79

Area of a Triangle with Inscribed Circle

12. When a circle is inscribed in a triangle, the sides of the triangle are tangents of the inscribed circle. In the triangle shown with its inscribed circle, we draw the three auxiliary radii to the points of tangency for the sides of the triangle. These radii separate the given triangle into triangles 1, 2, and 3. Where A_1, A_2, and A_3 represent the areas of the smaller triangles, the area A of the original triangle is given by the sum A = _____.

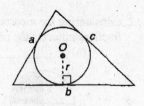

13. In this exercise, we continue to develop the concept found in Exercise 12. Because the altitude of each of the smaller triangles measures r (radius of the inscribed circle), the area A of the original given triangle is $A = \frac{1}{2}ra + \frac{1}{2}rb + \frac{1}{2}rc$.

Then $A = \frac{1}{2}r(a + b + c)$ or A = _____ , where P is the perimeter of the given triangle.

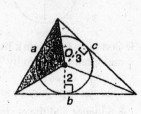

14. Using the formula $A = \frac{1}{2}rP$, find the area of the triangle whose sides have lengths 5 cm, 12 cm, and 13 cm and for which the radius of the inscribed circle has the length 2 cm. _____.

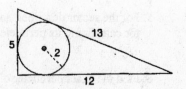

15. Heron's Formula, $A = \sqrt{s(s-a)(s-b)(s-c)}$, can be used to show that the area of a triangle whose sides have lengths of 13 ft, 14 ft, and 15 ft is 84 ft^2. Find the length of the radius of the circle that can be inscribed in this triangle. _____.

16. Find the length of the radius for the circle that can be inscribed in a right triangle whose legs measure 3 feet and 4 feet. _____.

17. In terms of a and b, find the length of the radius for the circle that can be inscribed in a right triangle whose legs have lengths a and b. _____.

80

Prisms

1. Considering its base and whether it is right or oblique, classify the type of prism shown.

   (a) _____ (b) _____ (c) _____

   _____

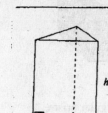 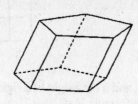 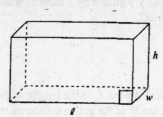

2. For the right triangular prism shown in Exercise 1(a), find its number of:
   (a) vertices _____ (b) lateral faces _____ (c) bases (base faces) _____
   (d) total faces _____ (e) lateral edges _____ (f) total edges _____

3. Find the lateral area of the right triangular prism shown in Exercise 1(a) given that the base edges measure 3 cm, 4 cm, and 5 cm respectively while the altitude measures 6 cm. _____.

4. Find the total area of the right triangular prism shown in Exercise 1(a) given that the base edges measure 3 cm, 4 cm, and 5 cm respectively while the altitude measures 6 cm. _____.
   [Hint: Each base is a right triangle.]

5. For the oblique pentagonal prism shown in Exercise 1(b), the area of each lateral face is 7.2 cm^2 and the area of each base measures 6.5 cm^2. For this prism, find its total area T. _____

6. For the right triangular prism shown at the right, the lateral area L is given by the formula $L = ah + bh + ch$ or $L = h(a + b + c)$. In turn, this formula becomes L = _____, where P is the perimeter of the base.

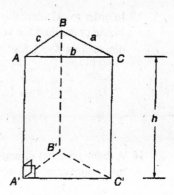

7. For the prism shown in Exercise 6, suppose that a = 5 in, b = 6 in, and c = 4 in. If the lateral area measures 150 in^2, use the formula $L = hP$ to find the length of the altitude of this prism. _____.

Volume

8. The cube has dimensions of length, width, and height. Each dimension measures 1 inch. In words, what unit of volume is represented by the cube? _____.

9. What symbol may be used for the expression *cubic inch*? _____

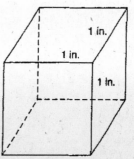

81

10. Complete (Volume Postulate): Correponding to every bounded solid is a unique positive number V known as the _____ of the solid.

11. The right rectangular prism (a box) has dimensions of 4 inches by 3 inches by 2 inches. What is the volume of the box? _____ .

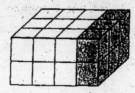

12. Complete: Where ℓ is the length, w the width, and h the height of a right rectangular prism, the volume V of the prism is determined by the formula V = _____ .

13. Complete: Where B represents the area of the base of a prism and h is the length of the altitude of the prism, the formula for the volume of the prism is V = _____ .

14. For the regular pentagonal prism shown, the apothem of the base measures 2.1 cm while each side measures 3 cm in length. Given that the length of the altitude is 5 cm, find:

   (a) the base area B for the prism. _____   [Recall: $A = \frac{1}{2}aP$.]

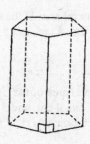

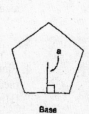

Base

   (b) the volume V of this pentagonal prism. _____ .

15. In order to calculate the volume of a solid figure, the dimensions of the solid must be measured in *like* units. Find the volume of a right rectangular prism (box) in *cubic inches* given that it has dimensions ℓ = 1 foot, w = 5 inches, and h = 6 inches. _____ .

16. A *cubic yard* has dimensions of 1 yard each. Given that 1 yard = 3 feet, how many cubic feet are in one cubic yard? _____ .

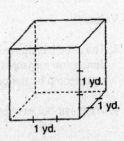

1 yd.

1 yd.

1 yd.

82

Pyramids

1. Considering its base and whether it is right or oblique, classify the type of pyramid shown.

(a) _____ (b) _____ (c) _____.
_____ _____ _____

Base is a square   Base is a square

2. For the right square pyramid shown in Exercise 1(a), find its number of:
   (a) vertices _____ (b) lateral faces _____ (c) bases (base faces) _____.
   (d) total faces _____ (e) lateral edges _____ (f) total edges _____.

3. For the regular square pyramid shown in Exercise 1(a), the length of the altitude is h = 4 inches while each side of the square base measures 6 inches. Find the length of the:
   (a) apothem of the square base _____ (b) the slant height of the pyramid _____.

4. For the regular square pyramid shown in Exercise 1(a), the length of the altitude is h = 8 cm while the length of the slant height is ℓ = 10 cm. Find the length of the:
   (a) apothem of the square base _____ (b) side of the square base _____.

5. For the regular pentagonal pyramid shown in Exercise 1(b), find the lateral area if it is known that the area of each triangular lateral face is 68 cm^2. _____.

6. For the pentagonal pyramid of Exercise 5, suppose that each side of the base measures 10 cm and that the apothem of the base measures 6.8 cm. Find the total area of the pentagonal prism. _____

   [Recall: The area of a regular polygon with apothem a and perimeter P is given by $A = \frac{1}{2}aP$.]

7. For the regular square pyramid shown, find the:
   (a) lateral area L _____ (b) total area T _____ .

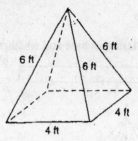

Volume

8. For a prism (Section 9.1) with base area B and length of altitude h, the volume can be found by the formula V = Bh. By experimentation, the volume of a pyramid with base area B and length of altitude h is determined by the formula V = _____ .

9. For a pyramid with base area $B = 12 \text{ in}^2$ and height $h = 6$ inches, find the volume. _____ .

10. Shown in the figure is a regular hexagonal pyramid. Each side of the base measures 4 inches while the length of the altitude is 12 inches. Find:
    (a) the area of the regular hexagonal base _____ .

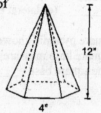 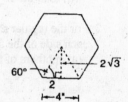

    (b) the volume of the regular hexagonal pyramid _____ .

11. Find the exact length of the slant height ℓ for the regular hexagonal pyramid shown in Exercise 10. _____ .

12. For a regular square pyramid (not shown), the length of the altitude is equal to the length of each side of the square base. If the volume of this pyramid is 72 cm^3, find the length of the altitude of the pyramid. _____ .

84

Cylinders

1. The bases of a circular cylinder are two _____ circles.

2. If the axis (line segment joining the centers of the circular bases) of a cylinder is perpendicular to each
   of the circular bases, then the cylinder is a(n) _____ circular cylinder.

Surface Area of a Cylinder

3. The lateral area of a cylinder is the limit of the surface areas of inscribed regular prisms with an
   increasing number of sides for their bases. Because the lateral area of a prism is given by L = hP
   (where h is the height and P is the perimeter of a base), the lateral area of the cylinder is given by
   L = _____ where h is the height and C is the circumference of a base. In terms of the height h and
   the length of radius r of the circular base, this formula becomes L = _____.

4. For the right circular cylinder shown, find the:
   (a) exact lateral area _____.

   (b) approximate lateral area to the nearest
       tenth of a square inch. _____.

5. For the right circular cylinder of Exercise 4, find the:
   (a) exact total area _____.

   (b) approximate total area to the nearest
       tenth of a square inch. _____.

Volume of a Cylinder

6. The volume of a right circular cylinder with length of altitude h and
   base area B is the limit of the volumes (V = Bh) of inscribed regular
   prisms with an increasing number of sides for their bases. In terms
   of the height h and the length of radius r of the circular base, the
   volume of the cylinder is V = _____.

7. For the right circular cylinder of Exercise 4, find the:
   (a) exact volume _____.

   (b) approximate volume to the nearest tenth
       of a cubic inch. _____.

Cones

8. Of the two circular cones shown, the first one is a(n) _____ circular cone while the second one is a(n) _____ circular cone.

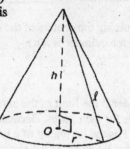

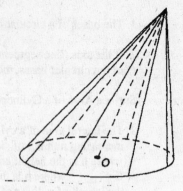

9. For the right circular cone of Exercise 8, find the length of the slant height ℓ if it is known that h = 6 cm and r = 4 cm. _____

Surface Area of a Right Circular Cone

10. Because the lateral area of a right circular cone is the limit of the lateral areas ($L = \frac{1}{2}\ell P$) of inscribed regular pyramids with increasing numbers of sides, the lateral area of the right circular cone with slant height length ℓ and base circumference becomes $L = \frac{1}{2}\ell C$ or $L = \frac{1}{2}\ell(2\pi r)$ or $L = $ _____.

11. For the right circular cone at the right, find the:

(a) exact lateral area _____.

(b) exact total area _____.

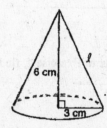

6 cm

3 cm

Volume of a Cone

12. Like the pyramid, the formula for the volume of a right circular cone with height h and area of base B can be written $V = \frac{1}{3}Bh$. Considering that the base of the cone is a circle of radius r, the formula for its volume is usually written $V = $ _____.

13. For the right circular cone of Exercise 11, find its:
    (a) exact volume. _____   (b) approximate volume to nearest tenth of cm^3. _____.

Solids of Revolution

14. What type of solid is generated when a rectangular region with dimensions of 2 ft and 5 ft is revolved about the 5 ft side as an axis? _____

2'

5'

15. What type of solid is generated when a right triangular region with legs of lengths 4 ft and 6 ft is revolved about the 6 ft leg as an axis? _____

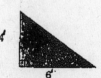

4

6'

86

Polyhedrons (Polyhedra)

1. When two planes intersect, the angle formed by two half-planes and the common edge (line of intersection of these planes) is known as a(n) _____ angle.

2. A _____ such as the one shown is a solid or region of space bounded by (enclosed by) plane regions.

3. Complete (Euler's Equation): The number of vertices V, number of edges E, and number of faces F of a polyhedron are related by the equation $V + F =$ _____.

4. For the pyramid shown in Exercise 6(a), Euler's Equation is verified by writing $4 + 4 = 6 + 2$. Verify Euler's Equation for the cube shown in Exercise 6(b). _____

5. While all prisms and pyramids are polyhedra, not all polyhedra are _____ or _____.

Regular Polyhedrons (Regular Polyhedra)

6. For the *regular* polyhedra (faces are congruent) shown, what are their names?
   (a) _____ (b) _____ (c) _____ (d) _____

7. Regular polyhedra are used as dice in many board games. Find the probability:
   (a) of rolling a "3 or larger" with a regular hexahedron (cube) whose faces are numbered 1, 2, 3, 4, 5, and 6? _____

   (b) of rolling a "number larger than 3" with a regular octahedron whose faces are numbered 1 through 8? _____

Surface Area of a Sphere

8. Complete (definition): A(n) _____ is the set of all points in space that lie at a fixed distance from a fixed point in space (known as the center of this space figure).

9. Complete: The surface area of a sphere of radius r is given by $S =$ _____.

10. For the sphere shown, $r = 1.2$ inches. Find:
    (a) the exact surface area of the sphere _____

    (b) the approximate surface area to the nearest tenth of a square inch. _____

Volume of a Sphere

11. Suppose that a sphere is subdivided into n pyramids, each with its vertex at the center of the sphere. Where B_1, B_2, B_3, and so on are the base areas of the pyramids, the volume of the sphere is the limit of the sum of volumes of the pyramids; that is, the volume of the sphere is the limit of $\frac{1}{3}B_1 h + \frac{1}{3}B_2 h + \frac{1}{3}B_3 h + \ldots + \frac{1}{3}B_n h$ or $\frac{1}{3}h(B_1 + B_2 + B_3 + \ldots + B_n)$. However, h → ____, the length of radius of the sphere while the sum of areas of the bases of the pyramids $B_1 + B_2 + B_3 + \ldots + B_n$ → _____, the surface area of the sphere. Then the volume of the sphere is $V = \frac{1}{3}r \cdot 4\pi r^2$ or V = _____.

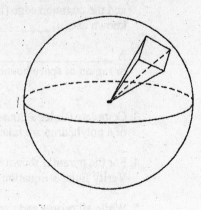

12. Use the formula $V = \frac{4}{3}\pi r^3$ to find the approximate volume of a sphere with the length of radius r = 3.7 inches. Answer to nearest tenth of cubic inch. _____. [Note: Use your calculator.]

13. A spherical storage tank is to be designed so that it has a capacity (volume) of 200 ft³. To the nearest tenth of a foot, find the length of radius of the sphere. _____.

14. Find the exact volume of a sphere whose exact surface area is 100π in². _____.

More Solids of Revolution

15. What type of solid is generated when the region bounded by a semicircle and its 12 cm diameter is revolved about the diameter? _____

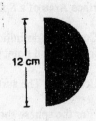

12 cm

16. For the sphere that was generated in Exercise 15, find its:
    (a) exact surface area _____.

    (b) exact volume _____.

The Rectangular Coordinate System

1. Consider the point (-5,4). Which number is the:
   (a) x-coordinate? _____ (b) y-coordinate? _____

2. For each point shown, give the coordinates (x,y).
   (a) A _____ (b) B _____ (c) C _____ (d) D _____

3. Name the quadrant or axis which identifies each point.
   (a) A _____ (b) B _____ (c) C _____ (d) D _____

4. In the rectangular coordinate system shown, plot and label
   each point: E(5,-6) and F(-3,0)

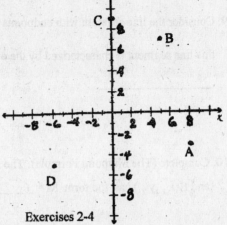

Exercises 2-4

Distance in the Rectangular Coordinate System

5. Find the distance between each pair of points:
   (a) (-3,4) and (5,4) _____ (b) (2.1,1.3) and (2.1,7.9) _____

6. Complete (The Distance Formula): The distance d between the two points (x_1, y_1) and (x_2, y_2)
   is given by the formula d = _____ .

7. Find the exact distance between each pair of points:
   (a) (-1,4) and (2,0) _____ (b) (0,3) and (2,9) _____

8. The equation of a line can generally be written in the form Ax + By = C. Complete the solution of the
   following problem and leave the solution in the form Ax + By = C.
   Find the equation of the line that describes the locus of points equidistant from A(5,-1) and B(-1,7).

   Solution: The following equation states that the point (x,y)
   is equidistant from A and B. All points of this form must lie
   on the perpendicular-bisector of $\overline{AB}$.

   $$\sqrt{(x-5)^2 + (y-(-1))^2} = \sqrt{(x-(-1))^2 + (y-7)^2} .$$
   Then

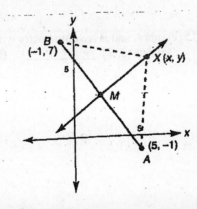

89

The Midpoint Formula

9. Consider the line segment with endpoints (x_1, y_1) and (x_2, y_2). The x-coordinate of the midpoint of this line segment is characterized by the expression $x_1 + \frac{1}{2}(x_2 - x_1)$. Simplify this expression.

   _____.

10. Complete (The Midpoint Formula): The midpoint M of the line segment joining the points $A(x_1, y_1)$ and $B(x_2, y_2)$ has the form $M = ($ _____ , _____ $)$.

11. Find the midpoint of the line segment that joins the points:
    (a) (-1,4) and (7,0) _____ (b) (a,b) and (c,d) _____.

12. Where C is the point (-3,5), the midpoint of $\overline{CD}$ is the point M(0,-4). Find the coordinates of point D. _____.

Symmetry in the Rectangular Coordinate System

13. Points A and B have symmetry with respect to the *y-axis*. Find B if:
    (a) A = (2,-5) _____ (b) A = (-3,5) _____.

14. Points A and B have symmetry with respect to the *x-axis*. Find B if:
    (a) A = (2,-5) _____ (b) A = (-3,5) _____.

15. Points A and B have symmetry with respect to the *origin*. Find B if:
    (a) A = (2,-5) _____ (b) A = (-3,5) _____.

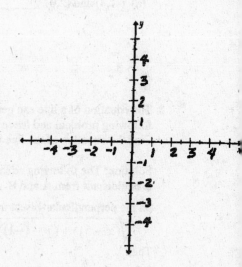

The Graph of an Equation

1. The graph of an equation is the set of points (x,y) in the rectangular coordinate system whose x and y coordinates satisfy (make true) the given _____.

2. The equation $Ax + By = C$ is called a linear equation because its graph is a(n) _____.

3. Complete the ordered pair solutions (points) that are on the graph of the equation $2x - 3y = 12$.
   (a) where x = 3 (3, ___)   (b) where y = -2 (___ , -2)

Intercepts

4. For the graph of a given equation, the x-intercept(s) is(are) the point(s) on the graph where y = ___.

5. Find the intercepts for the graph of the equation $3x + 4y = 24$.
   (a) x-intercept (___ , ___)   (b) y-intercept (___ , ___)

6. Using the intercepts found in Exercise 5, sketch the graph of the linear equation $3x + 4y = 24$ in the rectangular coordinate system provided.

7. Because the graph of the equation y = 2 is a horizontal line, it has only a(n) ___ - intercept.

8. Because the graph of the equation x = -5 is a vertical line, it has only a(n) ___ - intercept.

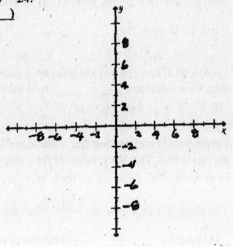

The Slope of a Line

9. Complete (Slope Formula): Where $x_1 \neq x_2$, the slope of the line that contains (x_1, y_1) and (x_2, y_2) is given by the formula m = _____.

10. Find the slope of the line that contains the points:
    (a) (-1,-4) and (2,6) _____   (b) (0,3) and (6,0) _____.

11. Find the slope of the line that is shown at the right. m = _____.

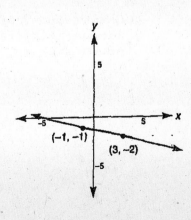

91

12. Find the slope of the line that contains the points:
    (a) (a,0) and (0,b) _____   (b) (c,d) and (e,f) _____.

Theorems Involving the Slope of a Line

13. Given three (or more) points on a line, the slope is *not changed* when:
    (a) the order of the selected points is _____.
    (b) two different points are _____.

14. Are the points A(2,-3), B(5,1), and C(-4,-11) collinear? _____.

15. In the rectangular coordinate system provided,
    draw the line through the point (1,-3) so that
    it has the slope m = $\dfrac{1}{4}$.

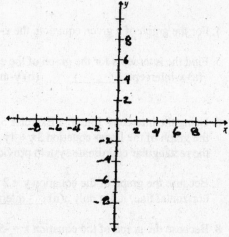

16. Complete: If two nonvertical lines are parallel,
    then their slopes are _____. In symbols,
    "If $\ell_1 \parallel \ell_2$, then $m_1 = m_2$."

17. Complete: If two lines (neither vertical nor horizontal) are
    perpendicular, then the product of their slopes equals____.
    In symbols, "If $\ell_1 \perp \ell_2$, then $m_1 \cdot m_2 = -1$."

18. The slope of line ℓ_1 is $m_1 = \dfrac{1}{3}$. Find the slope of line ℓ_2 if ℓ_1 and ℓ_2 are:

    (a) parallel _____   (b) perpendicular _____.

19. The slope of ℓ_1 is $m_1 = \dfrac{a}{b}$ where a ≠ 0 and b ≠ 0. Find the slope of line ℓ_2 if ℓ_1 and ℓ_2 are:

    (a) parallel _____   (b) perpendicular _____.

20. Given the vertices A(0,0), B(a,0), C(a + b,c), and D(b,c),
    we can use the Slope Formula to show that quadrilatreral
    ABCD is a(n) _____.

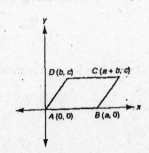

92

Formulas used in Analytic Proof

1. Complete (The Distance Formula): The distance d between the two points (x_1, y_1) and (x_2, y_2) in the rectangular coordinate system is given by the formula $d =$ _____.

2. Complete (The Midpoint Formula): The midpoint M of the line segment joining the points (x_1, y_1) and (x_2, y_2) is given by the formula $M =$ (_____ , _____).

3. Complete (The Slope Formula): Where $x_1 \neq x_2$, the slope of the line containing the points (x_1, y_1) and (x_2, y_2) is given by the formula $m =$ _____.

4. Complete: If two nonvertical lines ℓ_1 and ℓ_2 with slopes m_1 and m_2 are parallel, then their slopes are _____; in symbols, "If _____, then _____."

5. Complete: If two lines (neither vertical nor horizontal) ℓ_1 and ℓ_2 with slopes m_1 and m_2 are perpendicular, then the product of their slopes equals _____. "If $\ell_1 \perp \ell_2$, then _____."

6. Which formula(s)/relationship(s) would you use to show that:
   (a) two lines are parallel? _____.
   (b) two line segments are congruent? _____.
   (c) two line segments have the same midpoint? _____.

Applying Formulas and Relationships

7. Given the points A(2a,0) and B(0,2b), find:
   (a) the length of $\overline{AB}$. _____.

   (b) the midpoint of $\overline{AB}$. _____.

   (c) the slope of $\overline{AB}$. _____.

8. Given that the slope of ℓ_1 is $m_1 = \dfrac{c}{d}$ where $c \neq 0$ and $d \neq 0$, find the slope of:

   (a) ℓ_2 if $\ell_1 \parallel \ell_2$ _____.

   (b) ℓ_2 if $\ell_1 \perp \ell_2$ _____.

9. Does the point (a,b) lie on the graph of the equation $ax + by = a^2 + b^2$? _____.

Making a Drawing for an Analytic Proof

10. For an analytic proof, which point representation should be used . . . (a,b) or (2,3)? _____.

11. The figure used for an analytic proof must satisfy the _____ of the theorem without providing additional qualities; if the theorem calls for a rectangle, do not draw a square.

12. When placing the figure in the rectangular coordinate system, use as many _____ coordinates for points/vertices as possible.

13. It is convenient to use vertical and horizontal lines in a drawing because their _____ and perpendicular relationships are known.

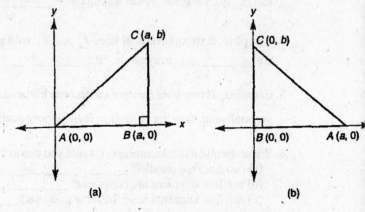

(a)   (b)

14. Which drawing, (a) or (b), is a better choice for a proof of a theorem of the form "In a right triangle, . . .?"

15. How would you label the coordinates of the vertices in exercise 14(b) if the conclusion of the theorem involves the midpoint of the hypotenuse? A = (___ , ___), B = (___ , ___), and C = (___ , ___)

Applications with Drawings

16. Given ▱ MNPQ, find the coordinates of vertex P in terms of a, c, and d. P = (_____ , _____).

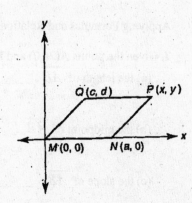

17. For ▱ MNPQ of Exercise 16, P = (a + c,d). Find:
   (a) the slope of both $\overline{MN}$ and $\overline{QP}$? _____.

   (b) the slope of both $\overline{MQ}$ and $\overline{NP}$? _____.

18. For ▱ ABCD, it is further known that ABCD is a rhombus. In terms of a, b, c, and d, what equation follows from the fact that:
   (a) AB = AD? _____.

   (b) $\overline{AC} \perp \overline{DB}$? _____.
   [Note: The diagonals are not shown.]

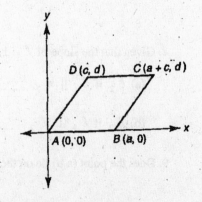

94

Figures used in Analytic Proof

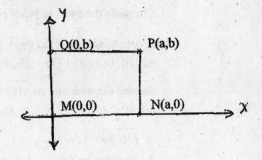

1. In the figure at the right, what is the relationship between:
   (a) either pair of opposite sides? _____.
   (b) any pair of adjacent sides? _____.

2. What type of quadrilateral is represented in
   the drawing for Exercise 1? _____.

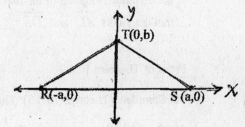

3. For $\triangle$ RST, how are sides $\overline{RT}$ and $\overline{ST}$ related?
   _____.

4. What type of triangle is shown in Exercise 3? _____

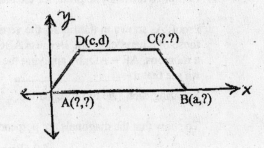

5. Given $\overline{AB} \parallel \overline{DC}$ and that trapezoid ABCD is isosceles,
   complete the coordinates for each vertex below:
   A = (____ , ____), B = (a, ____), and C = (____ , ____).

Proof of Theorem 10.4.1

6. Complete (Theorem 10.4.1): The line segment determined by the midpoints of two sides of a triangle is
   _____ to the third side.

7. Complete the analytic proof of the theorem stated in Exercise 5.

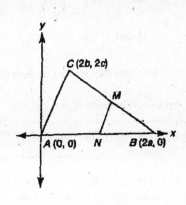

   Proof: As shown in the drawing, the vertices of $\triangle$ ABC are
   A(0,0), B(2a,0), and C(2b,2c). Where M is the midpoint of $\overline{CB}$,
   its coordinates are M = (____ , ____); likewise, with N as the
   midpoint of $\overline{AB}$, its coordinates are N = (____ , ____).
   Now the slope of $\overline{AC}$ is $m_{\overline{AC}}$ = _____ while the slope
   of $\overline{MN}$ is $m_{\overline{MN}}$ = _____. Because $m_{\overline{AC}} = m_{\overline{MN}}$,
   we know that _____.

Proof of Theorem 10.4.2

8. Complete (Theorem 10.4.2): The diagonals of a parallelogram _____ each other.

95

9. Complete the analytic proof of the theorem stated in Exercise 8.

Proof: As shown at the right, the vertices of ▱ ABCD are A(0,0), B(2a,0), C(2a + 2b,2c), and D(2b,2c). With $M_{\overline{AC}}$ naming the midpoint of $\overline{AC}$, the coordinates of this point are $M_{\overline{AC}} = ($ ____ , ____ $)$. Also, with $M_{\overline{DB}}$ as the midpoint of $\overline{DB}$, we have $M_{\overline{DB}} = ($ ____ , ____ $)$. Then (a + b,c) is the common midpoint of the two diagonals, which implies that diagonals $\overline{AC}$ and $\overline{DB}$ _____ each other.

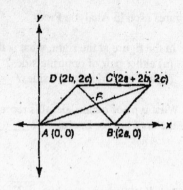

Proof of Theorem 10.4.3

10. Complete (Theorem 10.4.3): The diagonals of a rhombus are _____ .

11. Complete the analytic proof for the theorem in Exercise 10.

Proof: As shown in the figure, the vertices of ABCD are those of a parallelogram. Because ABCD further represents a rhombus, AB = AD; by applying the Distance Formula, we see that a = _____ .
Squaring each side of this equation, $a^2 = b^2 + c^2$ (*).

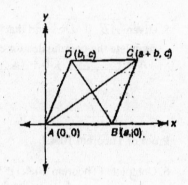

To show that the diagonals are perpendicular, we need slopes:
$$m_{\overline{AC}} = \underline{\hspace{2cm}} \quad \text{and} \quad m_{\overline{DB}} = \underline{\hspace{2cm}} .$$
Now the product of the slopes of the two diagonals is

$$m_{\overline{AC}} \cdot m_{\overline{DB}} = \frac{c}{a+b} \cdot \frac{-c}{a-b} = \frac{-c^2}{a^2-b^2} .$$ Using the starred

equation (*), $m_{\overline{AC}} \cdot m_{\overline{DB}} = \dfrac{-c^2}{(b^2+c^2)-b^2} = \underline{\hspace{1cm}}$ or ____ .

Because the product of their slopes is -1, we know that diagonals $\overline{AC}$ and $\overline{DB}$ are _____ .

Proof of Theorem 10.4.4

12. Complete (Theorem 10.4.4): if the diagonals of a parallelogram are equal in length, the parallelogram is a(n) _____ .

13. In order to sketch the proof of Theorem 9.4.4, we use the figure at the right. The fact that AC = DB leads to the conclusion that
$$a = 0 \text{ or } b = 0.$$
But coordinate a cannot possibly equal 0 because this would mean that A and B are the same point (likewise for D and C). Thus, it is necessary that _____ . With b = 0, vertices A, B, C, and D are those of a _____ .

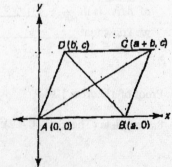

Linear Equations

1. Because the graphs of the equation $2x - 3y = 6$ and more generally $Ax + By = C$ are lines, such equations are known as _____ equations.

2 Because the graph of the equations $2x + 3y = 6$ and $y = -\dfrac{2}{3}x + 2$ are the same line, these

equations are known as _____ equations.

3. Are $2x - 3y = 6$ and $-6x + 9y = -18$ equivalent equations? _____.

The Slope-Intercept Form of a Line

4. The graph of a linear equation in slope-intercept form $y = mx + b$ is a line with slope _____ and y-intercept _____.

5. For the graph of the linear equation $y = 3x - 5$, what is:
   (a) the slope? m = _____   (b) the y-intercept? b = _____.

6. Change (transform) the equation $2x + 3y = 5$ into its equivalent slope-intercept form. y = _____.

7. Draw the graph of the equation $y = \dfrac{2}{3}x - 3$

   in the rectangular coordinate system provided.

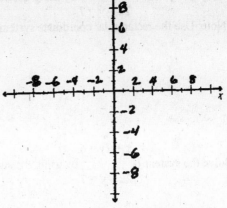

8. Find the general form $(Ax + By = C)$ for the line that has slope $m = -2$ and that contains the point $(0,5)$.

   _____.

The Point-Slope Form of a Line

9. The graph of a linear equation in the point-slope form $y - y_1 = m(x - x_1)$ is a line with slope _____ and which contains the point _____.

10. Find the general form $(Ax + By = C)$ for the line that has slope $m = \dfrac{1}{2}$ and that contains the

    point $(-2,5)$. _____.

11. Use the point-slope form to find the general form $(Ax + By = C)$ for the line that contains the points $(-2,5)$ and $(4,2)$. _____.

12. In the form $y = mx + b$, find the equation of the line that contains the points $(a,0)$ and $(0,b)$. _____.

Solving Systems of Equations

13. For a system of 2 linear equations, assume that the graphs intersect at a single point; in this case, the solution for the system is the ordered _____ represented by that point.

14. Solve the system $\{ \begin{matrix} x + 2y = 6 \\ 2x - y = 7 \end{matrix}$ by using the addition-subtraction method of algebra. (___ , ___)

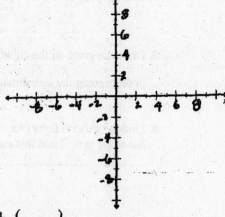

15. Solve the system $\{ \begin{matrix} x + 2y = 6 \\ 2x - y = 7 \end{matrix}$ by using geometry (graphing). (___ , ___)
   [Note: Use the rectangular coordinate system provided.]

16. Solve the system $\{ \begin{matrix} x + 2y = 6 \\ 2x - y = 7 \end{matrix}$ by using the substitution method of algebra. (___ , ___)

17. Solve the system $\{ \begin{matrix} y = ax + b \\ y = c \end{matrix}$ by using algebra. (___ , ___)

The Sine Ratio

1. For a right triangle, the sine ratio for an acute angle is found by dividing the length of the leg opposite the acute angle by the length of the _____.

2. In the right triangle (Δ ABC) for which m $\angle$ C $= 90^0$,

   $\sin \alpha = \dfrac{a}{c}$ while $\sin \beta =$ _____.

3. For the right triangle (Δ ABC), suppose that
   a = 3, b = 4, and c = 5. Find:
   (a) $\sin \alpha$ _____   (b) $\sin \beta$ _____.

4. For the right triangle (Δ ABC), suppose that
   a = 4 and b = 5. Find:
   (a) $\sin \alpha$ _____   (b) $\sin \beta$ _____.

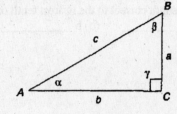

Exercises 2-4

5. The value $\sin 27^0$ can be approximated by drawing a right triangle with an angle that measures 27^0.

   Using a calculator, a more accurate approximation (to 4 decimal places) is $\sin 27^0 \approx$ _____.

Exact Values and Behavior of the Sine Ratio

6. Using any 45^0 - 45^0 - 90^0 right triangle, we can show that the exact value of $\sin 45^0$ is _____.
   [Note: Leave a square root radical in the answer.]

7. Using any 30^0 - 60^0 - 90^0 right triangle, we can show that the exact value of $\sin 30^0$ is _____ and
   that the exact value of $\sin 60^0$ is _____.
   [Note: Leave square root radicals in the answer as needed.]

8. $\sin 0^0 =$ _____ and $\sin 90^0 =$ _____.

9. As the measure θ of an acute angle of a right triangle increases, the value of $\sin \theta$ _____.

10. Without using a calculator, which is larger . . . $\sin 47^0$ or $\sin 53^0$? _____.

99

More About the Use of a Calculator

11. To use a calculator to find the sine ratio for an angle of a given degree measure, one must be sure that the calculator is placed in _____ mode.

12. Using a calculator, find each number correct to 4 decimal places:

(a) sin 13° _____   (b) sin 56.2° _____.

13. For the drawing provided, use the sine ratio to find the indicated length correct to the nearest tenth of a meter:

(a) a _____   (b) b _____.

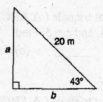

14. As found on a calculator, the notation "sin^{-1} (3/5)" is used to find the measure of the acute

_____ of a right triangle for which the ratio $\dfrac{opposite}{hypotenuse}$ equals $\dfrac{3}{5}$.

15. For the drawing provided, use the "sin^{-1} function" of your calculator to find the measure of the indicated angle correct to the nearest degree:

(a) α _____   (b) β _____.

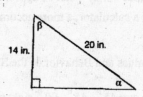

Applications: Angles of Elevation and of Depression

16. With respect to the ground, a kite is flying at an angle

of elevation of 67°. If there are 72 feet of string out to the kite, what is the altitude of the kite above the ground? Answer to the nearest whole number of feet. _____.

17. On a downhill slope, a skier moves down 10 feet for each 120 feet that she skies across the snow. To the nearest degree, what is the measure of the angle of depression of her downhill path? _____.

100

The Cosine Ratio

1. For a right triangle, the cosine ratio for an acute angle is found by dividing the length of the _____ that is adjacent to the right angle by the length of the hypotenuse.

2. In the right triangle ($\triangle ABC$) for which $m \angle C = 90^0$,

   $\sin \alpha = \dfrac{a}{c}$ while $\cos \alpha$ = _____.

3. For the right triangle ($\triangle ABC$), suppose that
   a = 3, b = 4, and c = 5. Find:
   (a) $\cos \alpha$ _____   (b) $\cos \beta$ _____.

4. For the right triangle ($\triangle ABC$), suppose that
   a = 5 and b = 7. Find:
   (a) $\cos \alpha$ _____   (b) $\cos \beta$ _____.

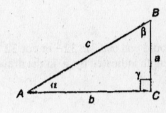

Exercises 2-4

5. In a right triangle whose acute angles measure 27^0 and 63^0, the value of $\cos 27^0$ can be shown to have the same value as sin _____.

Exact Values and Behavior of the Cosine Ratio

6. Using any 45^0 - 45^0 - 90^0 right triangle, we can show that the exact value of $\cos 45^0$ is _____.
   [Note: Leave a square root radical in the answer.]

7. Using any 30^0 - 60^0 - 90^0 right triangle, we can show that the exact value of $\cos 30^0$ is _____ and that the exact value of $\cos 60^0$ is _____.
   [Note: Leave square root radicals in the answer as needed.]

8. $\cos 0^0$ = _____ and $\cos 90^0$ = _____.

9. As the measure θ of an acute angle of a right triangle increases, the value of $\cos \theta$ _____.

10. Without using a calculator, which is larger . . . $\cos 35^0$ or $\cos 76^0$? _____.

More About the Use of a Calculator

11 Using a calculator, find each number correct to 4 decimal places:

   (a) $\cos 23^0$ _____   (b) $\cos 62.7^0$ _____.

12. For the drawing provided, use the cosine ratio to find the indicated length correct to the nearest tenth of a unit:

   (a) c _____   (b) d _____.

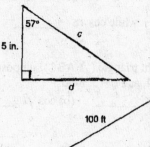

13. Which should you use, $\sin 32^0$ or $\cos 32^0$, if you were to find the length indicated by a in the drawing? _____.

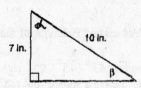

14. When found on a calculator, the notation "$\cos^{-1}(5/8)$" is used to find the _____ of the acute angle of a right triangle for which the ratio $\dfrac{adjacent}{hypotenuse}$ equals $\dfrac{5}{8}$.

15. For the drawing provided, use the "$\cos^{-1}$ function" of your calculator to find the measure of the indicated angle correct to the nearest degree:

   (a) α _____   (b) β _____.

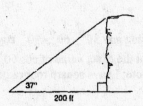

Applications: Angles of Elevation and of Depression

16. At a point 200 feet from the base of a cliff, the top of the cliff is seen through an angle of elevation of 37^0. To the nearest foot, find the distance from the point on the ground to the top of the cliff? _____.

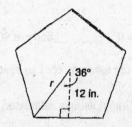

17. Use the drawing provided to find the length of the radius of a regular pentagon whose apothem measures 12 inches. Give the answer to the nearest tenth of an inch. _____.

The Tangent Ratio

1. While the cosine ratio for an acute angle of a right triangle takes the form $\dfrac{adjacent}{hypotenuse}$,

   the tangent ratio has the form _____.

2. In the right triangle ($\triangle ABC$) for which $m\angle C = 90^0$,

   $\sin \alpha = \dfrac{a}{c}$, $\cos \alpha = \dfrac{b}{c}$, and $\tan \alpha =$ _____.

3. For the right triangle ($\triangle ABC$), suppose that
   $a = 8$ and $b = 15$. Find:
   (a) $\tan \alpha$ _____   (b) $\tan \beta$ _____.

4. For the right triangle ($\triangle ABC$), suppose that
   $a = 6$ and $c = 10$. Find:
   (a) $\tan \alpha$ _____   (b) $\tan \beta$ _____.

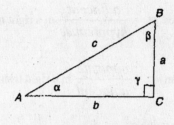

Exercises 2-4

Exact Values and Behavior of the Tangent Ratio

5. Using any 45^0 - 45^0 - 90^0 right triangle, we can show that the exact value of $\tan 45^0$ is _____.

6. Using any 30^0 - 60^0 - 90^0 right triangle, we can show that the exact value of $\tan 30^0$ is _____ and
   that the exact value of $\tan 60^0$ is _____.
   [Note: Leave square root radicals in the answer as needed.]

7. $\tan 0^0 =$ _____ while $\tan 90^0$ is _____.

8. As the measure θ of an acute angle of a right triangle increases, the value of $\tan \theta$ _____.

Using a Calculator

9. Using a calculator, find each number correct to 4 decimal places:
   (a) $\tan 37^0$ _____   (b) $\tan 89.5^0$ _____.

10. Without using a calculator, which is larger . . . $\tan 37^0$ or $\tan 53^0$? _____.

11. For the drawing provided, use the tangent ratio
    to find the length y correct to the nearest tenth
    of a unit. _____.

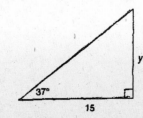

103

12. Use the calculator and the "tan^{-1} function" to
find the value of θ to the nearest degree. _____ .

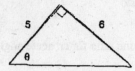

Other Trigonometric Ratios

13. While $\sin \alpha = \dfrac{opposite}{hypotenuse}$ for acute angle α in a right triangle, its reciprocal is the cosecant
of α . The ratio for csc α is _____ .

14. While $\cos \alpha = \dfrac{adjacent}{hypotenuse}$ in a right triangle, its reciprocal is the secant of α ; sec α = _____ .

15. While $\tan \alpha = \dfrac{opposite}{adjacent}$ in a right triangle, its reciprocal is the cotangent of α ; cot α = _____ .

16. Find the value of:

(a) sec α if $\cos \alpha = \dfrac{2}{5}$. _____ (b) csc β if $\sin \beta = 0.25$ _____ .

17. Using a calculator, find each number correct to 4 decimal places:

(a) cot 13^0 _____ (b) sec 52^0 _____ .

Applications

18. Which ratio would you use to find the length a . . .
sin 52^0 , cos 52^0 , or tan 52^0? _____ .

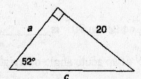

19. The topmost point of a lookout tower is seen from a point
270 feet from its base. If the angle of elevation is 37^0 , find
the height of the tower to the nearest foot. _____ .

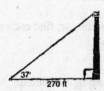

104

The Area of a Triangle

1. Complete: The area of a triangle equals one-half the product of the lengths of two sides of the triangle and the sine of their _____ angle.

2. Fill the blanks to complete the proof of the theorem found in Exercise 1.

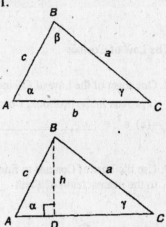

    Given: Acute triangle ABC

    Prove: $A = \dfrac{1}{2}$ bc sin α

    Proof: The area of Δ ABC is given by $A = \dfrac{1}{2}$ bh, where

    h is the length of the _____ that corresponds to the base whose length is b.

    In rt. triangle ABD, sin α = _____; in turn, h = _____.

    Substitution for h in the equation $A = \dfrac{1}{2}$ bh leads to the

    formula $A = \dfrac{1}{2}$ b(_____) or A = _____.

3. Complete: Two equivalent formulas for the area of Δ ABC (in Exercise 2) are

    $A = \dfrac{1}{2}$ ac sin β and A = _____.

4. To the nearest tenth of a square meter, find the area of the triangle shown at the right. _____

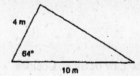

The Law of Sines

5. Because the area of a triangle is unique, we know that the area of Δ ABC in Exercise 2 is unique.

    That is, $\dfrac{1}{2}$ bc sin α = $\dfrac{1}{2}$ ac sin β = $\dfrac{1}{2}$ ab sin γ

    Multiplying each member of the extended equation above by 2, we have:
            bc sin α = ac sin _____ = _____.

    Dividing each member of the equation bc sin α = ac sin β = ab sin γ by the product abc,

    we obtain the Law of Sines _____.

6. Use the Law of Sines, $\dfrac{\sin \alpha}{a} = \dfrac{\sin \beta}{b} = \dfrac{\sin \gamma}{c}$ to find the length x

    correct to the nearest tenth of a foot. _____.

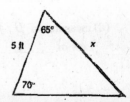

105

7. Use the Law of Sines to find the measure α
   to the nearest degree. _____

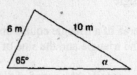

The Law of Cosines

8. One form of the Law of Cosines, namely, $c^2 = a^2 + b^2 - 2ab \cos \gamma$
   is proven in the textbook. Complete two additional equivalent forms:
   (a) $a^2 =$ _____   (b) $b^2 =$ _____

9. Use the Law of Cosines to find the length x
   to the nearest tenth of a unit. _____

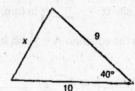

10. Use the Law of Cosines to find the measure β
    to the nearest degree. _____

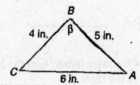

When to use the Law of Sines/Cosines

11. For the drawing provided, state whether you would use the
    Law of Sines or the Law of Cosines to find the missing measure,
    and state the *form* of that law that you would use.
    (a) measures of a, b, and c are known; find α .
        Law of _____ ; _____

    (b) measures of β, b, and c are known; find γ .
        Law of _____ ; _____

    (c) measures of β, a, and c are known; find b.
        Law of _____ ; _____

    (d) measures of β, γ, and c are known; find a.
        Law of _____ ; _____

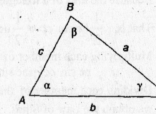

106

Section 1.1
2. a, b
4. a. F b. T c. T d. T
6. a. F b. T c. T d. F
7. a. T b. T
8. AM and MB have the same length
10. Katrina has seen snow.
11. No conclusion
13. Left
15. Triangle ABC has 2 equal angles.

Section 1.2
3. angle; △
5. 3.7 cm
7. 14.8
8. 6
10. a. 35 b. 21
11. a. 61 b. 34 c. 32.5
18. $x = 6$
19. $x = 10$; $y = 7$

Section 1.3
3. triangle
5. line
6. positive
7. AX + XB = AB
10. intersect
14. parallel
16. line
17. 6
18. 29 and 5

Section 1.4
2. m∠ABC
5. obtuse
8. adjacent
10. complementary
12. vertical
15. 71
17. $x = 25$
19. $x = 10$
20. $x = 23$

Section 1.5
2. a. $2x = 10$ b. $x = 5$
3. a. Distributive
   b. Substitution
   c. Transitive
6. R1 Given
   R2 Distributive
   R3 Substitution
   R4 Addition of Eq.
   R5 Div. (or Mult.) of Eq.

Section 1.5
8. a. Segment Addition Post.
   b. Addition Prop. of Eq.
10. R1 Given
    R2 Add. Prop. Of Eq.
    R3 Segment Add. Post.
    R4 Substitution

Section 1.6
1. Perpendicular
4. a. Reflexive b. Symmetric
   c. Transitive
6. R, S, and T
7. Trans.
8. Symmetric
10. a. 1 b. ∞
14. a. 142 b. x c. $180 - x$ d. 12

Section 1.7
1. P; Q
4 a. theorem b. hypothesis
   c. hypothesis d. conclusion
   e. Given; Prove
6. If Q, then P.
7. No
10. 62
11. R1. Given
    R2. Def'n supp. angles
    S3. m∠3 + m∠2
    R4. Subtraction
    S5. ∠1 ≅ ∠3
    R5. Def'n congruent angles
12. a. 52 b. 131 c. 45
14. 28

Section 2.1
3. parallel
4. 3, 4, 5, 6
6. 3 and 6; or 4 and 5
9. supplementary
10. S1. a ∥ b; trans. K
    R1. Given
    R2. corresponding
    R3. vertical
    S4. ∠3 ≅ ∠6
12. a. $x = 42$
    b. $y = 10$
    c. $z = 16\frac{2}{3}$
    d. $w = 20$

Section 2.2
2. Contrapositive
3. ~P
4. Pablo does not live
   in Guadalajara.
6. a and c
7. a, b, and c
9. vertical angles;
   ∠1 ≅ ∠2; false;
   ∠s 1 and 2 are not vert. ∠s

Section 2.3
2. lines are parallel
4. supplementary
6. coplanar
9. $x = 13$
10. $x = 20$
12. None
13. c and d
16. R1 Given
    R2 If 2 ∥ lines are cut by a
       trans, the corr. ∠s are ≅.
    R3 Vertical angles are ≅.
    S4 ∠3 ≅ ∠4
    S6 ℓ ∥ n

Section 2.4
2. isosceles
6. obtuse
8. 180
11. 60
12. complementary
14. exterior
16. $x = \dfrac{80}{6}$ or $x = 13\dfrac{1}{3}$
17. 139
18. m∠1 = $x + 180 - y$

Section 2.5
3. $\overline{AC}$ and $\overline{BD}$
5. 20
6. $n - 2$
9. 1800
10. $I = \dfrac{(n-2) \cdot 180}{n}$
11. a. 60 b. 108 c. 144
13. 360
15. polygram

Section 2.6
1. horizontal
4. a. vert. b. horiz.
    c. both d. vert.
6. a, c, and d
7. no
9. congruent
14. no

Section 3.1
1. a. $\angle$D b. vertex F c. $\overline{AC}$
3. a. $\overline{UV}$ b. $\angle$U
5. $\overline{MP} \cong \overline{MP}$
7. SSS
10. AAA
12. a. ASA b. SAS
13. R1 Given
    R2 Def'n of bisect
    R3 Given
    S4 $\triangle$ AMC $\cong$ $\triangle$ BMD
    R4 SSS

Section 3.2
3. $\angle$D
5. R1 Given
    S2 $\angle$TWZ $\cong$ $\angle$VWZ
    R3 Given
    S4 $\overline{WZ} \cong \overline{WZ}$
    R5 SAS
    S6 $\overline{TZ} \cong \overline{VZ}$
6. S7 Z is the midpoint of $\overline{TV}$
    R7 Def'n of midpoint
10. legs; hypotenuse
12. $a^2 + b^2$
14. a. 5 b. 6 c. $\sqrt{41}$ d. $\sqrt{95}$

Section 3.3
3. $\angle$BAC
4. altitude
7. legs; base
9. $\angle$P
10. R1 Given
    S2 $\overline{BD}$ bisects $\angle$ABC
    R3 Def'n of bisector of angle
    S4 $\overline{BD} \cong \overline{BD}$
    S5 $\triangle$ ABD $\cong$ $\triangle$ CBD
    R5 SAS

Section 3.3
13. $\angle$T $\cong$ $\angle$U
15. a. 68 b. 44
17. a. 52 b. 64
18. 33
19. x = 9

Section 3.4
1. R1 Given
    S2 $\overline{DE} \cong \overline{TR}$
    R3 SSS
    S4 $\angle$B $\cong$ $\angle$S
    R4 CPCTC
3. R1 Given
    S2 $\overline{YW} \cong \overline{YW}$
    R3 SSS
    S4 $\angle$XYW $\cong$ $\angle$ZYW
    S5 $\overline{YW}$ bisects $\angle$XYZ
6. a. 120 b. Extend one side of 60 degree angle to form an adjacent angle that measures 120.
7. a. 135 b. Extend one side of 45 degree angle to form an adjacent angle that measures 135.

Section 3.5
4. AB; BC
6. nonadjacent
10. a. $\angle$B b. $\angle$A
     c. m$\angle$B > m$\angle$C > m$\angle$A
12. a. 60 b. $\overline{BC}$ c. $\overline{AC}$
     d. BC > AB > AC
15. greater than
16. sum; difference
17. a. Yes b. No c. No

Section 4.1
1. parallel
2. R1 Given
    R3 alternate interior
    S4 $\overline{AD} \parallel \overline{BC}$
    S5 $\angle$3 $\cong$ $\angle$4
    R6 Identity
    R7 ASA
6. supplementary
9. a. $10\frac{2}{3}$ b. $11\frac{2}{3}$ c. $16\frac{1}{3}$
10. a. 50 b. 60 c. 120

Section 4.1
11. a. 65 b. 75 c. 105
13. a. 6.3 b. 8.5
14. $\overline{QN}$
17. N 30°E

Section 4.2
3. bisect
5. adjacent
6. angles
9. 10
10. S1 Kite ABCD with
     $\overline{BC} \cong \overline{CD}, \overline{AD} \cong \overline{AB}$
     R3 Identity
     R4 SSS
     S5 $\angle$B $\cong$ $\angle$D
     R5 CPCTC
12. a. 6.4 b. 10.6
     c. 2x + 6 d. x = 9
14. a. 10.2 b. 14.4
     c. 13.8 d. 38.4
15. parallelogram

Section 4.3
3. right
4. S1 Rect MNPQ with
     diagonals $\overline{MP}$ and $\overline{NQ}$
     R1 Given
     R3 Opp. sides of a $\square$ are $\cong$.
     R4 Identity
     R5 All $\angle$s of a rect. are rt. $\angle$s.
     S6 $\angle$NMQ $\cong$ $\angle$PQM
     R7 SAS
     S8 $\overline{MP} \cong \overline{NQ}$
     R8 CPCTC
6. congruent
7. congruent; perpendicular
10. 22.8 cm
13. 5 feet
14. 13 cm

Section 4.4
2. a. $\overline{HL}$, $\overline{JK}$ b. $\overline{HJ}$, $\overline{LK}$
    c. $\angle$H and $\angle$L (or $\angle$s J and K)
3. a. No b. Yes
4. median
6. altitude
9. bases
10. a. 57 b. 108 c. 18

Section 4.4
12. a. 10 b. 10 c. 17 d. 24
15. S1 Trap. RSTV with

$\overline{RS} \parallel \overline{VT}$ and $\overline{RV} \cong \overline{ST}$

R3 Def'n of parallelogram
R4 congruent
R6 opposite
R7 corresponding
S8 $\angle V \cong \angle T$
R8 Trans. Prop. Of Congruence
17. 4.2

Section 5.1
1. a. 3 to 4 b. 3:5 c. 1:3
2. a. 16 mi/gal b. 14 cents/pencil
4. a. $x = \dfrac{35}{9}$ b. x = 3 or x = -3

  c. x = 8   d. x = 7 or x = -5
5. geometric mean
7. 203 miles
8. 36^0 and 54^0
9. 30^0, 60^0, and 90^0
12. 6
13. x = 12.5 and y = 15

Section 5.2
1. a. True b. False
3. a. congruent b. proportional
4. a. No b. Yes c. Yes d. No
5. a. 27 b. 92 c. 61

7. a. $\dfrac{40}{7}$ b. $\dfrac{80}{7}$

9. a. 30 b. 60
11. 3.0 inches
13. 4 or 12

Section 5.3
2. a. Yes b. No c. No
3. S1 $\overline{AB} \parallel \overline{DE}$
  R1 Given
  R2 alternate int. $\angle$ s are $\cong$
  R3 Vertical $\angle$ s are $\cong$.
  S4 $\triangle ABC \sim \triangle EDC$
  R4 AA
4 a. sides b. angles
6. Means-Extremes
8. a. SAS ~ b. SSS ~
9. congruent; similar

Section 5.3
10. a. AA b. SAS ~ c. SSS ~
11. a. 80 b. 35 c. 65

Section 5.4
2. Identity; right
4. a. $\sqrt{52} = 2\sqrt{13}$ b. 5
6. a. Yes b. No c. Yes
7. 50 feet
9. $c^2 = a^2 + b^2$
10. a. Yes b. No
  c. No (not integral) d. Yes
11. a. (3, 4, 5) b. (5, 12, 13)
13. a. Obtuse b. Right c. Acute

Section 5.5
2. $a\sqrt{2}$

4. a. 6; $6\sqrt{2}$ b. 8; 8

5. a. $3\sqrt{2}$; 6 b. $2\sqrt{2}$; $2\sqrt{2}$
8. 2a

10. shorter; $\sqrt{3}$

11. a. 16; $8\sqrt{3}$ b. 6: $6\sqrt{3}$

12. a. 10: 20 b. 6; $4\sqrt{3}$

Section 5.6
1. $\dfrac{DE}{EF}$

2. 4.25
5. 7.5
6. x = 2
8. 8.4

9. $\dfrac{4}{6} = \dfrac{2}{3}$

12. 4.5
13. 4

14. $\dfrac{16}{15}$

Section 6.1
2. a. 5 b. 10 c. $5\sqrt{2}$
4. central; minor; major; semicircle
6. a. 58 b. 122 c. 122 d. 29
8. 3

11. chords; inscribed; $\overset{\frown}{AC}$

Section 6.1
14. a. 27 b. 38
15. a. 58 b. 36.7
16. right

Section 6.2
1. tangent; tangency
3. inscribed; chords; quadrilateral
5. supplementary
7. vertical
9. a. 56 b. 124
12. a. 41 b. 49 c. 90 d. 98
13. difference
15. a. 39 b. 107
16. a. 39 b. 38
17. 55

Section 6.3
1. chord; arc
4. a. 8 b. 16
5. internally
6. a. 3 b. 2 c. 0
7. congruent
9. a. 5.2 b. 9.5 c. 11.4 d. 12.3
11. product
13. a. 5.25 b. 3 or 8
14. 16
16. a. 12 b. 6

Section 6.4
1. tangent
3. tangent
6. a. m$\angle$1 > m$\angle$2

  b. m$\overset{\frown}{AB}$ > m$\overset{\frown}{DC}$
9. a. $\overline{EF}$ b. $\overline{AB}$
11. a. 4 b. 8 c. 3 d. 6
12. AB > CD

14. m$\overset{\frown}{AB}$ > m$\overset{\frown}{CD}$
16. MN > PQ

Section 7.1
1. locus
2. the circle with center P and radius length of 1 inch.
3. sphere
5. the bisector of the angle
7. the perpendicular-bisector of the line segment.
9. Each marked angle is inscribed in a semicircle.

Selected Answers for the Interactive Companion for Elementary Geometry for College Students, 5e

Section 7.2
1. concurrent
4. E; incenter
5. equidistant
6. inscribe
9. F; circumcenter
10. vertices
11. circumscribe
14. N; orthocenter
15. exterior
18. C; centroid
19. a. 8 b. 4
20. a. 15 b. 5
21. a. 9 b. 12 c. 8
   d. 4 e. $\sqrt{97}$

Section 7.3
2. a. No b. Yes
   c. Yes d. No
3. All
5. a. Yes b. No
   c. Yes d. Yes
6. All
7. 108
8. 45
12. radius
14. 10
15. $3.4\sqrt{2}$ cm
16. apothem
18. $3\sqrt{3}$ ft
19. $4\sqrt{3}$ cm
20. $24\sqrt{5}$ in

Section 8.1
3. equal
5. 20.3 cm^2
7. 12 cm^2
8. 60 in^2
9. 144 in^2
12. 28 cm^2
14. 40 units^2
15. $\frac{1}{2}$ bh
17. 24 units^2
19. 24 cm^2
20. a. 10 cm b. 4.8 cm

Section 8.2
4. a. 16 cm b. 20 ft c. 32 in
7. 6 cm^2
8. $6\sqrt{6}$ in^2
10. $\frac{1}{2}$ h(b$_1$ + b$_2$)
11. 36 in^2
12. 7 cm
16. 40 ft^2
17. 20 in
20. $\frac{9}{4}$
21. $\frac{1}{9}$

Section 8.3
2. 81 in^2
3. 100 cm^2
6. $25\sqrt{3}$ in^2
7. $36\sqrt{3}$ cm^2
11. $A = \frac{1}{2}aP = \frac{1}{2}(\frac{s}{2\sqrt{3}})(3s)$
    $= \frac{s^2}{4} \cdot \frac{3}{\sqrt{3}} = \frac{s^2}{4}\sqrt{3}$
12. $150\sqrt{3}$ cm^2
14. 219.76 ft^2
15. 720 cm^2

Section 8.4
2. π ; $\frac{22}{7}$
4. a. 14π in b. 44 in
5. a. 7 cm b. 14 cm
8. a. 4π cm b. 12.6 cm
10. a. 10; 0
11. 90
13. 16π cm^2
14. 141.0 in^2
15. 38.5 in^2

Section 8.4
16. 18.9 m^2
17. 38.8 cm^2

Section 8.5
2. a. $\frac{1}{4}$ b. $\frac{1}{3}$
3. $\frac{m}{360} \cdot \pi r^2$
4. 71.0 in^2
6. $(12 + 4\pi)$ ft
8. A$_{sector}$ - A$_\Delta$
9. $(36\pi - 72)$ in^2
10. $(12\sqrt{2} + 6\pi)$ in
11. $(108\pi + 72)$ in^2
13. $\frac{1}{2}$ rP
14. 30 cm^2
15. 4 ft
16. 1 foot
17. $\frac{ab}{a+b+\sqrt{a^2+b^2}}$

Section 9.1
3. 72 cm^2
4. 84 cm^2
5. 49 cm^2
7. 10 in
11. 24 in^3
14. a. 15.75 cm^2 b. 78.75 cm^3
15. 360 in^3
16. 27 ft^3

Section 9.2
3. a. 3 in b. 5 in
4. a. 6 cm b. 12 cm
5. 340 cm^2
6. 510 cm^2
7. a. $32\sqrt{2}$ ft^2
   b. $(32\sqrt{2} + 16)$ ft^2
9. 24 in^3
10. a. $24\sqrt{3}$ in^2 b. $96\sqrt{3}$ in^3

Section 9.2

11. $\sqrt{156}$ in or $2\sqrt{39}$ in

12. 6 cm

Section 9.3

4. a. 120π in^2 b. 377.0 in^2

5. a. 170π in^2 b. 534.1 in^2

7. a. 300π in^3 b. 942.5 in^3

9. $\sqrt{52}$ cm or $2\sqrt{13}$ cm

11. a. $9\pi\sqrt{5}$ cm^2

  b. $(9\pi\sqrt{5}+9\pi)$ cm^2

13. a. 18π cm^3 b. 56.5 cm^3

14. right circular cylinder

15. right circular cone

Section 9.4

4. $8+6=12+2$

6. a. tetrahedron b. cube
  c. octahedron d. dodecahedron

7. a. $\dfrac{2}{3}$ b. $\dfrac{5}{8}$

10. a. 5.76π in^2 b. 18.1 in^2

12. 212.2 in^3

13. 3.6 ft

14. $\dfrac{500\pi}{3}$ in^3

16. a. 144π cm^2 b. 288π cm^3

Section 10.1

2. a. (8,-3) b. (5,7)
  c. (0,9) d. (-6,-5)

3. a. IV b. I c. y-axis d. III

5. a. 8 b. 6.6

7. a. 5 b. $\sqrt{40}$ or $2\sqrt{10}$

8. $3x-4y=-6$ or $-3x+4y=6$

11. a. (3,2) b. $\left(\dfrac{a+c}{2},\dfrac{b+d}{2}\right)$

12. (3,-13)

14. a. (2,5) b. (-3,-5)

15. a. (-2,5) b. (3,-5)

Section 10.2

3. a. -2 b. 3

5. a. (8,0) b. (0,6)

10. a. $\dfrac{10}{3}$ b. $-\dfrac{1}{2}$

11. $-\dfrac{1}{4}$

12. a. $-\dfrac{b}{a}$ b. $\dfrac{f-d}{e-c}$

18. a. $\dfrac{1}{3}$ b. -3

19. a. $\dfrac{a}{b}$ b. $-\dfrac{b}{a}$

Section 10.3

1. $\sqrt{(x_2-x_1)^2+(y_2-y_1)^2}$

6. a. Slope b. Distance
  c. Midpoint

7. a. $2\sqrt{a^2+b^2}$ b. (a,b) c. $-\dfrac{b}{a}$

8. a. $\dfrac{c}{d}$ b. $-\dfrac{d}{c}$

13. parallel

15. (2a,0), (0,0), and (0,2b)

16. (a+c,d)

18. a. $a=\sqrt{c^2+d^2}$

  b. $\dfrac{d}{a+c}\cdot\dfrac{-d}{a-c}=-1$

Section 10.4

1. a. parallel b. perpendicular

5. (0,0), (a,0), and (a-c,d)

6. parallel

8. bisect

10. perpendicular

13. b = 0; rectangle

Section 10.5

5. a. 3 b. -5

6. $-\dfrac{2}{3}x+\dfrac{5}{3}$

Section 10.5

8. $2x+y=5$ or $-2x-y=-5$

10. $x-2y=-12$ or $-x+2y=12$

11. $x+2y=8$ or $-x-2y=-8$

14. (4,1)

17. $\left(\dfrac{c-b}{a},c\right)$

Section 11.1

3. a. $\dfrac{3}{5}$ b. $\dfrac{4}{5}$

4. a. $\dfrac{4\sqrt{41}}{41}$ b. $\dfrac{5\sqrt{41}}{41}$

5. 0.4540

6. $\dfrac{\sqrt{2}}{2}$

7. $\dfrac{1}{2}$, $\dfrac{\sqrt{3}}{2}$

10. sin 53

12. a. 0.2250 b. 0.8310

13. a. 13.6 m b. 14.6 m

15. a. 44 b. 46

16. 66 ft

17. 5 degrees

Section 11.2

3. a. $\dfrac{4}{5}$ b. $\dfrac{3}{5}$

4. a. $\dfrac{7\sqrt{74}}{74}$ b. $\dfrac{5\sqrt{74}}{74}$

5. 63^0

7. $\dfrac{\sqrt{3}}{2}$, $\dfrac{1}{2}$

10. cos 35^0

11. a. 0.9205 b. 0.4586

12. a. 9.2 b. 7.7

15. a. 46^0 b. 44^0

16. 250 feet

17. 14.8 in

111

Section 11.3

3. a. $\dfrac{8}{15}$ b. $\dfrac{15}{8}$

4. a. $\dfrac{3}{4}$ b. $\dfrac{4}{3}$

6. a. $\dfrac{\sqrt{3}}{3}$ b. $\sqrt{3}$

9. a. 0.7536 b. 114.5887

11. 11.3

12. 50^{0}

16. a. 2.5 b. 4

17. a. 4.3315 b. 1.6243

19. 203 feet

Section 11.4

3. $\dfrac{1}{2}$ ab sin γ

4. 18.0 m^{2}

6. 6.6 feet

7. 33^{0}

9. 6.6 units

10. 83^{0}

11. a. Cosines; $a^{2} = b^{2} + c^{2} - 2bc \cos \alpha$

    b. Sines; $\dfrac{\sin \gamma}{c} = \dfrac{\sin \beta}{b}$

    c. Cosines; $b^{2} = a^{2} + c^{2} - 2ac \cos \beta$

    d. Sines; $\dfrac{\sin \alpha}{a} = \dfrac{\sin \gamma}{c}$ where $\alpha = 180 - (\beta + \gamma)$

Chapter 1 Line and Angle Relationships

SECTION 1.1: Sets, Statements, and Reasoning

1. a. Not a statement.

  b. Statement; true

  c. Statement; true

  d. Statement; false

5. Conditional

9. Simple

13. H: The diagonals of a parallelogram are perpendicular.

  C: The parallelogram is a rhombus.

17. First, write the statement in "If, then" form. If a figure is a square, then it is a rectangle.

  H: A figure is a square.

  C: It is a rectangle.

21. True

25. Induction

29. Intuition

33. Angle 1 looks equal in measure to angle 2.

37. *A Prisoner of Society* might be nominated for an Academy Award.

41. Angles 1 and 2 are complementary.

45. None

49. Marilyn is a happy person.

53. Not valid

SECTION 1.2: Informal Geometry and Measurement

1. $AB < CD$

5. One; none

9. Yes; no; yes

13. Yes; no

17. a. 3

  b. $2\frac{1}{2}$

21. Congruent; congruent

25. No

29. Congruent

33. $\overline{AB}$

37. $x + x + 3 = 21$
  $2x = 18$
  $x = 9$

41. 71

45. $32.7 \div 3 = 10.9$

49. N 22° E

SECTION 1.3: Early Definitions and Postulates

1. AC

5. $\frac{1}{2}$ m $\cdot$ 3.28 ft/m $= 1.64$ feet

9. a. A-C-D

  b. A, B, C or B, C, D or A, B, D

13. a. m and t

  b. m and $\overrightarrow{AD}$ or $\overrightarrow{AD}$ and t

17. $2x + 1 + 3x = 6x - 4$
  $5x + 3 = 6x - 4$
  $-1x = -7$
  $x = 7;\ AB = 38$

21. a.

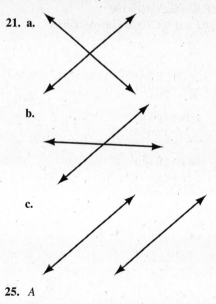

  b.

  c.

25. A

29. Given: $\overline{AB}$ and $\overline{CD}$ as shown $(AB > CD)$
Construct $\overline{MN}$ on line l so that
$MN = AB + CD$

A ●————————● B

C ●————● D

◄————●————————●————————●————————► l
    M   N

33. **a.** No

    b. Yes

    c. No

    d. Yes

37. Nothing

SECTION 1.4: Angles and Their Relationships

1. **a.** Acute

    b. Right

    c. Obtuse

5. Adjacent

9. **a.** Yes

    b. No

13. $m\angle FAC + m\angle CAD = 180$
$\angle FAC$ and $\angle CAD$ are supplementary.

17. $42°$

23. $\quad\quad x + y = 2x - 2y$
$x + y + 2x - 2y = 64$

$\quad\quad -1x + 3y = 0$
$\quad\quad\ \ 3x - 1y = 64$

$\quad\quad -5x + 2y = 10$
$\quad\quad\ \ \underline{5x + 2y = 78}$
$\quad\quad\quad\quad\ 22y = 88$
$\quad\quad\quad\quad\quad\ y = 4;\ x = 14$

27. $\quad x + y = 180$
$\quad x = 24 + 2y$

$\quad x + y = 180$
$\quad x - 2y = 24$

$\quad -2x + 2y = 360$
$\quad\ \ \underline{x - 2y = 24}$
$\quad\ 3x\quad\quad = 384$
$\quad\quad\quad x = 128;\ y = 52$

31. $x - 92 + (92 - 53) = 90$
$\quad\ x - 92 + 39 = 90$
$\quad\quad\quad x - 53 = 90$
$\quad\quad\quad\quad\quad x = 143$

35. Given: Obtuse $\angle MRP$
Construct: Rays RS, RT, and RU so that $\angle MRP$ is divided into $4 \cong$ angles.

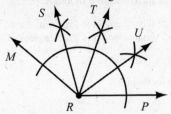

39. It appears that the two sides opposite $\angle$ s A and B are congruent.

45. $\quad 90 + x + x = 360$
$\quad\quad\quad\ 2x = 270$
$\quad\quad\quad\quad x = 135$

SECTION 1.5: Introduction to Geometric Proof

1. Division Property of Equality or Multiplication Property of Equality

5. Multiplication Property of Equality

9. Angle-Addition Property

13. $\overrightarrow{EG}$ bisects $\angle DEF$

17. $2x = 10$

21. $6x - 3 = 27$

25. **1.** $2(x + 3) - 7 = 11$

    2. $2x + 6 - 7 = 11$

    3. $2x - 1 = 11$

    4. $2x = 12$

    5. $x = 6$

29. **1.** Given

    2. If an angle is bisected, then the two angles formed are equal in measure.

    3. Angle-Addition Postulate

    4. Substitution

    5. Distribution Property

    6. Multiplication Property of Equality

33. $5 \cdot x + 5 \cdot y = 5(x + y)$

SECTION 1.6: Relationships: Perpendicular Lines

1. **1.** Given

    2. If 2 $\angle$ s are $\cong$, then they are equal in measure.

    3. Angle-Addition Postulate

    4. Addition Property of Equality

    5. Substitution

    6. If 2 $\angle$ s are = in measure, then they are $\cong$.

5. Given: Point N on line s.
Construct: Line m through N so that $m \perp s$.

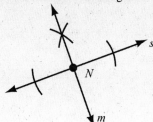

9. Given: Triangle ABC
Construct: The perpendicular bisectors of each side, $\overline{AB}$, $\overline{AC}$, and $\overline{BC}$.

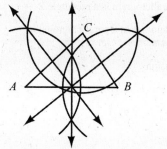

13. No; Yes; No

17. No; Yes; Yes

23. **1.** $M - N - P - Q$ on $\overline{MQ}$ **1.** Given

    2. $MN + NQ = MQ$ **2.** Segment-Addition Postulate

    3. $NP + PQ = NQ$ **3.** Segment-Addition Postulate

    4. $MN + NP + PQ = MQ$ **4.** Substitution

27. In space, there are an infinite number of lines which perpendicularly bisect a given line segment at its midpoint.

SECTION 1.7: The Formal Proof of a Theorem

1. H: A line segment is bisected.

    C: Each of the equal segments has half the length of the original segment.

5. H: Each is a right angle.

    C: Two angles are congruent.

11. Given: $\overleftrightarrow{AB} \perp \overleftrightarrow{CD}$
Prove: $\angle AEC$ is a right angle.

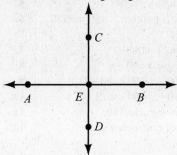

15. Given: Lines l and m
Prove: $\angle 1 \cong \angle 2$ and $\angle 3 \cong \angle 4$

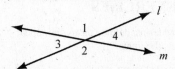

19. $m\angle 1 = m\angle 3$
    $3x + 10 = 4x - 30$
       $x = 40; \ m\angle 1 = 130°$

25. **1.** Given

    2. If 2 $\angle$ s are comp., then the sum of their measures is 90.

    3. Substitution

    4. Subtraction Property of Equality

    5. If 2 $\angle$ s are = in measure, then they are $\cong$.

29. 1. Given

  2. $\angle ABC$ is a right $\angle$.

  3. The measure of a rt. $\angle = 90$.

  4. Angle-Addition Postulate

  6. $\angle 1$ is comp. to $\angle 2$.

33. The supplement of an acute angle is obtuse.
   Given: $\angle 1$ is supp to $\angle 2$
     $\angle 2$ is an acute $\angle$
   Prove: $\angle 1$ is an obtuse $\angle$

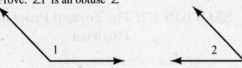

STATEMENTS	REASONS
1. $\angle 1$ is supp to $\angle 2$	**1.** Given
2. $m\angle 1 + m\angle 2 = 180$	**2.** If 2 $\angle$s are supp., the sum of their measures is 180.
3. $\angle 2$ is an acute $\angle$	**3.** Given
4. $m\angle 2 = x$ where $0 < x < 90$	**4.** The measure of an acute $\angle$ is between 0 and 90.
5. $m\angle 1 + x = 180$	**5.** Substitution (#4 into #3)
6. x is positive $\therefore m\angle 1 < \angle 180$	**6.** If $a + p_1 = b$ and p_1 is positive, then $a < b$.
7. $m\angle 1 = 180 - x$	**7.** Substitution Prop of Eq. (#5)
8. $-x < 0 < 90 - x$	**8.** Subtraction Prop of Ineq. (#4)
9. $90 - x < 90 < 180 - x$	**9.** Addition Prop. or Ineq. (#8)
10. $90 - x < 90 < m\angle 1$	**10.** Substitution (#7 into #9)
11. $90 < m\angle 1 < 180$	**11.** Transitive Prop. of Ineq (#6 & #10)
12. $\angle 1$ is an obtuse $\angle$	**12.** If the measure of an angle is between 90 and 180, then the $\angle$ is obtuse.

REVIEW EXERCISES

1. Undefined terms, defined terms, axioms or postulates, theorems.

2. Induction, deduction, intuition

3. 1. Names the term being defined.

  2. Places the term into a set or category.

  3. Distinguishes the term from other terms in the same category.

  4. Reversible

4. Intuition

5. Induction

6. Deduction

7. H: The diagonals of a trapezoid are equal in length.

  C: The trapezoid is isosceles.

8. H: The parallelogram is a rectangle.

  C: The diagonals of a parallelogram are congruent.

9. No conclusion

10. Jody Smithers has a college degree.

11. Angle A is a right angle.

12. C

13. $\angle RST$, $\angle S$, more than $90°$.

14. Diagonals are $\perp$ and they bisect each other.

15.

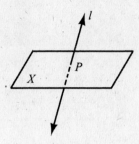

16.

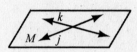

17.

18. a. Obtuse   **b.** Right

19. a. Acute   **b.** Reflex

20. $2x + 15 = 3x - 2$
$17 = x$
$x = 17;\ m\angle ABC = 98°$

21. $2x + 5 + 3x - 4 = 86$
$5x + 1 = 86$
$5x = 85$
$x = 17;\ m\angle DBC = 47°$

22. $3x - 1 = 4x - 5$
$4 = x$
$x = 4;\ AB = 22$

23. $4x - 4 + 5x + 2 = 25$
$9x - 2 = 25$
$9x = 27$
$x = 3;\ MB = 17$

24. $2 \cdot CD = BC$
$2(2x + 5) = x + 28$
$4x + 10 = x + 28$
$3x = 18$
$x = 6;\ AC = BC = 6 + 28 = 34$

25. $7x - 21 = 3x + 7$
$4x = 28$
$x = 7$
$m\angle 3 = 49 - 21 = 28°$
$\therefore m\angle FMH = 180 - 28 - 152°$

26. $4x + 1 + x + 4 = 180$
$5x + 5 = 180$
$5x = 175$
$x = 35$
$m\angle 4 = 35 + 4 = 39°$

27. a. Point M

b. $\angle JMH$

c. $\overline{MJ}$

d. $\overleftrightarrow{KH}$

28. $2x - 6 + 3(2x - 6) = 90$
$2x - 6 + 6x - 18 = 90$
$8x - 24 = 90$
$8x = 114$
$x = 14\frac{1}{4}$

$m\angle EFH = 3(2x - 6) = 3\left(28\frac{1}{2} - 6\right)$
$= 3 \cdot 22\frac{1}{2}$
$= 67\frac{1}{2}°$

29. $x + (40 + 4x) = 180$
$5x + 40 = 180$
$5x = 140$
$x = 28$

30. a. $2x + 3 + 3x - 2 + x + 7 = 6x + 8$

b. $6x + 8 = 32$
$6x = 24$
$x = 4$

c. $2x + 3 = 2(4) + 3 = 11$
$3x - 2 = 3(4) - 2 = 10$
$x + 7 = 4 + 7 = 11$

31. The measure of angle 3 is less than 50.

32. The four foot board is 48 inches. Subtract 6 inches on each end leaving 36 inches.
$4(n - 1) = 36$
$4n - 4 = 36$
$4n = 40$
$n = 10$
$\therefore$ 10 pegs will fit on the board.

33. S

34. S

35. A

36. S

37. N

38. **2.** $\angle 4 \cong \angle P$

     3. $\angle 1 \cong \angle 4$

     4. If 2 $\angle$ s are $\cong$, then their measures are =.

     5. Given

     6. $m\angle 2 = m\angle 3$

     7. $m\angle 1 + m\angle 2 = m\angle 4 + m\angle 3$

     8. Angle-Addition Postulate

     9. Substitution

    10. $\angle TVP \cong \angle MVP$

39. Given: $\overline{KF} \perp \overline{FH}$
            $\angle JHK$ is a right $\angle$
    Prove: $\angle KFH \cong \angle JHF$

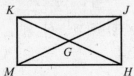

STATEMENTS	REASONS
1. $\overline{KF} \perp \overline{FH}$	**1.** Given
2. $\angle KFH$ is a right $\angle$	**2.** If 2 segments are $\perp$, then they form a right $\angle$.
3. $\angle JHF$ is a right $\angle$	**3.** Given
4. $\angle KFH \cong \angle JHF$	**4.** Any two right $\angle$s are $\cong$.

40. Given: $\overline{KH} \cong \overline{FJ}$

      G is the midpoint of both $\overline{KH}$ and $\overline{FJ}$

    Prove: $\overline{KG} \cong \overline{GJ}$

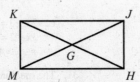

STATEMENTS	REASONS
1. $\overline{KH} \cong \overline{FJ}$ G is the midpoint of both $\overline{KH}$ and $\overline{FJ}$	**1.** Given
2. $\overline{KG} \cong \overline{GJ}$	**2.** If 2 segments are $\cong$, then their midpoints separate these segments into 4 $\cong$ segments.

41. Given: $\overline{KF} \perp \overline{FH}$
  Prove: $\angle KFH$ is comp to $\angle JHF$

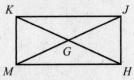

STATEMENTS	REASONS
1. $\overline{KF} \perp \overline{FH}$	1. Given
2. $\angle KFH$ is comp. to $\angle JFH$	2. If the exterior sides of 2 adjacent $\angle$s form $\perp$ rays, then these $\angle$s are comp.

42. Given: $\angle 1$ is comp. to $\angle M$
  $\angle 2$ is comp. to $\angle M$
  Prove: $\angle 1 \cong \angle 2$

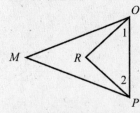

STATEMENTS	REASONS
1. $\angle 1$ is comp. to $\angle M$	1. Given
2. $\angle 2$ is comp. to $\angle M$	2. Given
3. $\angle 1 \cong \angle 2$	3. If 2 s are comp. to the same , then these angles are $\cong$.

43. Given: $\angle MOP \cong \angle MPO$
  $\overrightarrow{OR}$ bisects $\angle MOP$
  $\overrightarrow{PR}$ bisects $\angle MPO$
  Prove: $\angle 1 \cong \angle 2$

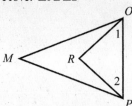

STATEMENTS	REASONS
1. $\angle MOP \cong \angle MPO$	1. Given
2. $\overrightarrow{OR}$ bisects $\angle MOP$ $\overline{PR}$ bisects $\angle MPO$	2. Given
3. $\angle 1 \cong \angle 2$	3. If 2 s are $\cong$, then their bisectors separate these $\angle$s into four $\cong$ $\angle$s.

44. Given: ∠4 ≅ ∠6
Prove: ∠5 ≅ ∠6

STATEMENTS	REASONS
1. ∠4 ≅ ∠6	**1.** Given
2. ∠4 ≅ ∠5	**2.** If 2 angles are vertical ∠s then they are ≅.
3. ∠5 ≅ ∠6	**3.** Transitive Prop.

45. Given: Figure as shown
Prove: ∠4 is supp. to ∠2

STATEMENTS	REASONS
1. Figure as shown	**1.** Given
2. ∠4 is supp. to ∠2	**2.** If the exterior sides of 2 adjacent ∠s form a line, then the ∠s are supp.

46. Given: ∠3 is supp. to ∠5
∠4 is supp. to ∠6
Prove: ∠3 ≅ ∠6

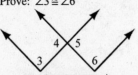

STATEMENTS	REASONS
1. ∠3 is supp to ∠5 ∠4 is supp to ∠6	**1.** Given
2.	If 2 lines intersect, the vertical angles **2.** formed are ≅.
3. ∠3 ≅ ∠6	**3.** If 2 ∠s are supp to congruent angles, then these angles are ≅.

47. Given: $\overline{VP}$
Construct: $\overline{VW}$ such that $VW = 4 \cdot VP$

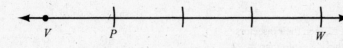

48. Construct a 135° angle.

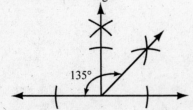

49. Given: Triangle *PQR*
    Construct: The three angle bisectors.

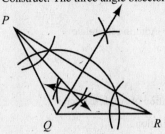

It appears that the three angle bisectors meet at one point inside the triangle.

50. Given: $\overline{AB}$, $\overline{BC}$, and $\angle B$ as shown
    Construct: Triangle *ABC*

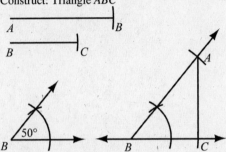

51. Given: m$\angle B = 50°$
    Construct: An angle whose measure is 20°.

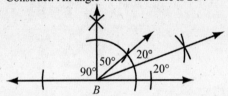

52. m$\angle 2 = 270°$

CHAPTER TEST

1. Induction

2. $\angle CBA$ or $\angle B$

3. $\overline{AP} + \overline{PB} = \overline{AB}$

4. **a.** point

   b. line

5. **a.** Right

   b. Obtuse

6. **a.** Supplementary

   b. Congruent

7. m$\angle MNP =$ m$\angle PNQ$

8. **a.** right

   b. supplementary

9. Kianna will develop reasoning skills.

10. $3.2 + 7.2 = 10.4$ in.

11. **a.** $x + x + 5 = 27$
    $2x + 5 = 27$
    $2x = 22$
    $x = 11$

    b. $x + 5 = 11 + 5 = 16$

12. m$\angle 4 = 35°$

13. **a.** $x + 2x - 3 = 69$
    $3x - 3 = 69$
    $3x = 72$
    $x = 24$

    b. m$\angle 4 = 2(24) - 3 = 45°$

14. **a.** m$\angle 2 = 137°$

    b. m$\angle 2 = 43°$

15. **a.** $2x - 3 = 3x - 28$
    $x = 25$

    b. m$\angle 1 = 3(25) - 28 = 47°$

16. **a.** $2x - 3 + 6x - 1 = 180$
    $8x - 4 = 180$
    $8x = 184$
    $x = 23$

    b. m$\angle 2 = 6(23) - 1 = 137°$

17. $x + y = 90$

18.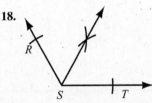

19.

20. **1.** Given

    2. Segment-Addition Postulate

    3. Segment-Addition Postulate

    4. Substitution

21. **1.** $2x - 3 = 17$

    2. $2x = 20$

    3. $x = 10$

22. **1.** Given

    2. 90°

    3. Angle-Addition Postulate

    4. 90°

    5. Given

    6. Definition of Angle-Bisector

    7. Substitution

    8. $m\angle 1 = 45°$

23. 108°

Chapter 2: Parallel Lines

SECTION 2.1: The Parallel Postulate and Special Angles

1. a. 108°

   b. 72°

5. a. No

   b. Yes

   c. No

9. a. $m\angle 3 = 87°$; $\angle 3$ is vertical to $\angle 2$.

   b. $m\angle 6 = 87°$; $\angle 6$ corresponds to $\angle 2$.

   c. $m\angle 1 = 93°$; $\angle 1$ is supplementary to $\angle 2$.

   d. $m\angle 7 = 87°$; $\angle 7$ corresponds to $\angle 3$.

13. a. $m\angle 2 = 68°$; $\angle 2$ is supp. to $\angle 1$.

   b. $m\angle 4 = 112°$; $\angle 4$ is vertical to $\angle 1$.

   c. $m\angle 5 = 112°$; $\angle 5$ is an alternate interior $\angle$ to $\angle 4$.

   d. $m\angle MOQ = 34°$

     $m\angle MON = m\angle 2 = 68°$

     $m\angle MOQ = \dfrac{1}{2}$ of $m\angle MON = 34°$.

17. Angles 3 and 5 are supp. because they are interior angles on the same side of the transversal. Angles 5 and 6 are also supp. This leads to a system of 2 equations with 2 variables.

   $(6x + y) + (8x + 2y) = 180$

   $(8x + 2y) + (4x + 7y) = 180$

   Simplifying yields,

   $14x + 3y = 180$

   $12x + 9y = 180$

   Dividing the 2nd equation by –3 gives

   $14x + 3y = 180$

   $-4x - 3y = -60$

   Addition gives

   $10x = 120$

   $x = 12$

   Using $14x + 3y = 180$ and $x = 12$ we get

   $14(12) + 3y = 180$

   $168 + 3y = 180$

   $3y = 12$

   $y = 4$

   $m\angle 6 = 4(12) + 7(4) = 76°$

   $\therefore m\angle 7$ also $= 76°$.

21. Given: $\overleftrightarrow{CE} \parallel \overleftrightarrow{DF}$; trans. $\overleftrightarrow{AB}$

   $\angle JHK$ is a right $\angle$

   Prove: $\angle KFH \cong \angle JHF$

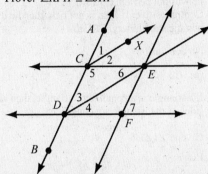

STATEMENTS	REASONS
1. $\overleftrightarrow{CE} \parallel \overleftrightarrow{DF}$; trans. $\overleftrightarrow{AB}$	1. Given
2. $\angle ACE \cong \angle ADF$	2. If 2 ∥ lines are cut by a trans., then the corresponding $\angle$s are $\cong$.
3. $\overrightarrow{CX}$ bisects $\angle ACE$ $\overrightarrow{DE}$ bisects $\angle CDF$	3. Given
4. $\angle 1 \cong \angle 3$	4. If two $\angle$s are $\cong$, then their bisectors separate these $\angle$s into four $\cong$ $\angle$s.

27. a. $\angle 4 \cong \angle 2$ and $\angle 5 \cong \angle 3$

   b. 180°

   c. 180°

31. No

35. Given: Triangle MNQ with obtuse $\angle MNQ$

   Construct: $\overline{MR} \perp \overline{NQ}$

   (Hint: Extend $\overline{NQ}$)

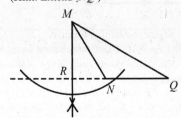

SECTION 2.2: Indirect Proof

1. If Juan wins the state lottery, then he will be rich.
   Converse: If Juan is rich, then he won the state lottery. FALSE.
   Inverse: If Juan does not win the state lottery, then he will not be rich. FALSE.
   Contrapositive: If Juan is not rich, then he did not win the state lottery. TRUE.

5. No conclusion.

9. (a) (b) (e)

13. If a triangle is an equilateral triangle, then all sides of the triangle are congruent.

19. Given: $\angle 1 \not\cong \angle 5$

   Prove: $r \not\parallel s$

   Proof:
   Assume that $r \parallel s$. If they are $\parallel$, then $\angle 1 \cong \angle 5$ because they are corresponding angles. But this contradicts the Given information. Therefore, our assumption is false and $r \not\parallel s$.

23. Assume that the angles are vertical angles. If they are vertical angles, then they are congruent. But this contradicts the hypothesis that the two angles are not congruent. Hence, our assumption must be false and the angles are not vertical angles.

27. Given: M is a midpoint of $\overline{AB}$.

   ●————————●————————●
   A M B

   Prove: M is the only midpoint of $\overline{AB}$.
   Proof: If M is a midpoint of $\overline{AB}$, then
   $AM = \frac{1}{2} \cdot AB$. Assume that N is also a midpoint
   of $\overline{AB}$ so that $AN = \frac{1}{2} \cdot AB$. By substitution
   $AM = AN$.

   ●————●——●————————●
   A N M B

   By the Segment-Addition Postulate,
   $AM = AN + NM$. Using substitution again,
   $AN + NM = AN$. Subtracting gives $NM = 0$.
   But this contradicts the Ruler Postulate which states that the measure of a line segment is a positive number. Therefore, our assumption is wrong and M is the only midpoint for $\overline{AB}$.

SECTION 2.3: Proving Lines Parallel

1. $l \parallel m$

5. $l \not\parallel m$ (Sum = 181°)

9. None

13. None

17. 1. Given

    2. If 2 $\angle$s are comp. to the same $\angle$, then they are $\cong$.

    3. $\overline{BC} \parallel \overline{DE}$

21. Given: $\overrightarrow{DE}$ bisects $\angle CDA$
    $\angle 3 \cong \angle 1$
    Prove: $\overline{ED} \parallel \overline{AB}$

    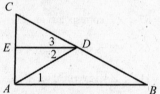

STATEMENTS	REASONS
1. $\overleftrightarrow{CE} \parallel \overleftrightarrow{DF}$; trans. $\overleftrightarrow{AB}$	1. Given
2. $\angle ACE \cong \angle ADF$	2. If 2 $\parallel$ lines are cut by a trans., then the corresponding $\angle$s are $\cong$.
3. $\overrightarrow{CX}$ bisects $\angle ACE$ $\overrightarrow{DE}$ bisects $\angle CDF$	3. Given
4. $\angle 1 \cong \angle 3$	4. If two $\angle$s are $\cong$, then their bisectors separate these $\angle$s into four $\cong$ $\angle$s.

27. $x^2 - 9 = x(x-1)$
$x^2 - 9 = x^2 - x$
$x = 9$

31. If two lines are cut by a transversal so that the alternate exterior angles are congruent, then these lines are parallel.
Given: Lines l and m and trans t;
$\angle 1 \cong \angle 2$
Prove: $l \parallel m$

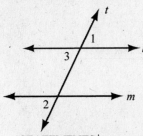

STATEMENTS	REASONS
1. Lines l and m and trans. t $\angle 1 \cong \angle 2$	1. Given
2. $\angle 1 \cong \angle 3$	2. If 2 lines intersect, the vertical $\angle$s formed are $\cong$.
3. $\angle 2 \cong \angle 3$	3. Transitive for $\cong$.
4. $l \parallel m$	4. If lines are cut by a trans. so that the corresponding angles are $\cong$, then these lines are $\parallel$.

35. Given: Line l and P not on l
Construct: The line through $P \parallel l$

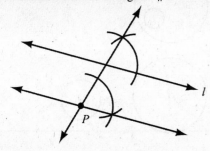

SECTION 2.4: The Angles of a Triangle

1. $\text{m}\angle C = 75°$

5. a. Underdetermined

b. Determined

c. Overdetermined

9. a. Equiangular Δ

b. Right Δ

13. $\text{m}\angle 1 = 122°$; $\text{m}\angle 2 = 58°$; $\text{m}\angle 5 = 72°$

17. 35°

23. 360°

29. $x + 4y = 180$
$2y + 2x - y - 40 = 180$

$x + 4y = 180$ (Multiply by -2)
$2x + y = 220$

$-2x - 8y = -360$
$\underline{2x + y = 220}$
$-7y = -140$
$y = 20$
$x + 4(20) = 180$
$x + 100 = 180$
$x = 100$
$\text{m}\angle 2 = 80°$; $\text{m}\angle 3 = 40°$ $\therefore \text{m}\angle 5 = 60°$

33. a.

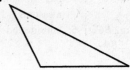

b. It is not possible to draw an equilateral right triangle.

37. $x + 2x + 33 = 180$
$$3x = 147$$
$$x = 49$$

m$\angle N = 49°$; m$\angle P = 98°$

41. 75°

45. The measure of an exterior angle of a triangle equals the sum of measures of the two nonadjacent interior angles.
Given: $\triangle ABC$ with ext. $\angle BCD$
Prove: m$\angle BCD$ = m$\angle A$ + m$\angle B$

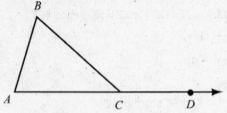

STATEMENTS	REASONS
1. $\triangle ABC$ with ext. $\angle BCD$	1. Given
2. m$\angle A$ + m$\angle B$ + m$\angle BCA$ = 180	2. The sum of the measures of the $\angle$s in a $\triangle$ is 180.
3. $\angle BCA$ is supp. to $\angle BCD$	3. If the exterior sides of two adjacent $\angle$s form a straight line, then the angles are supp.
4. m$\angle BCA$ + m$\angle BCD$ = 180	4. If two $\angle$s are supp., then the sum of their measures is 180.
5. m$\angle BCA$ + m$\angle BCD$ = m$\angle A$ + m$\angle B$ + m$\angle BCA$	5. Substitution
6. m$\angle BCD$ = m$\angle A$ + m$\angle B$	6. Subtraction Prop. of Equality

49. $2b = $ m$\angle M + 2a$
(m$\angle RPM$ = m$\angle M$ + m$\angle MNP$)
$\therefore$ m$\angle M = 2b - 2a$
$b = 42 + a$
(m$\angle QPR$ = m$\angle Q$ + m$\angle QNP$)
m$\angle M = 2(42 + a) - 2a$ (Substitution)
m$\angle M = 84 - 2a + 2a$
m$\angle M = 84°$

SECTION 2.5: Convex Polygons

1. Increase

For #5 use $D = \dfrac{n(n-3)}{2}$

5. a. 5

b. 35

For #9 use $I = \dfrac{180(n-2)}{n}$

9. a. 90°

b. 150°

For #13 use $S = 180(n-2)$

13. a. 7

b. 9

For #17 use $n = \dfrac{360}{E}$

17. a. 15

b. 20

21.

25.

29. Given: Quad. *RSTV* with diagonals
    $\overline{RT}$ and $\overline{SV}$ intersecting at *W*
    Prove: $m\angle 1 + m\angle 2 = m\angle 3 + m\angle 4$

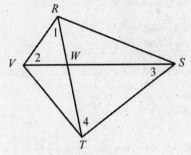

STATEMENTS	REASONS
1. Quad *RSTV* with diagonals $\overline{RT}$ and $\overline{SV}$ intersecting at *W*.	1. Given
2. $m\angle RWS = m\angle 1 + m\angle 2$	2. The measure of an ext. $\angle$ of a $\triangle$ equals the sum of the measures of the non-adjacent interior $\angle$s of the $\triangle$.
3. $m\angle RWS = m\angle 3 + m\angle 4$	3. Same as (2)
4. $m\angle 1 + m\angle 2 = m\angle 3 + m\angle 4$	4. Substitution

33. $36°$

For #37 use $I = \dfrac{180(n-2)}{n}$

37. $150°$

41. $221°$

SECTION 2.6: Symmetry and Transformations

1. M, T, X

5. (a), (c)

9. MOM

13. a.
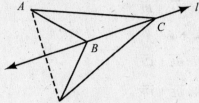

b.
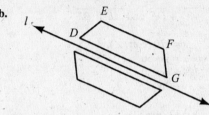

17. WHIM

21. WOW

25. 62,365 kilowatt hours

29. (b), (c)

CHAPTER REVIEW

1. a. $\overline{BC} \parallel \overline{AD}$

   b. $\overline{AB} \parallel \overline{CD}$

2. $m\angle 3 = 110°$

3. $2x + 17 = 5x - 94$
   $111 = 3x$
   $37 = x$

4. $m\angle A = 50°$ (corresponds to $\angle DCE$)
   $\therefore m\angle BCA = 55$
   $m\angle BCD = 75$ and $m\angle D = 75$
   $m\angle DEF = 50 + 75 = 125$

5. $130 + 2x + y = 180$
   $150 + 2x - y = 180$

   $2x + y = 50$
   $\underline{2x - y = 30}$
   $4x = 80$
   $x = 20$

   $130 + 2(20) + y = 180$
   $130 + 40 + y = 180$
   $y = 10$

6. $2x + 15 = x + 45$
   $x = 30$

   $3y + 30 + 45 = 180$
   $3y + 75 = 180$
   $3y = 105$
   $y = 35$

7. $\overline{AE} \parallel \overline{BF}$

8. None

9. $\overline{BE} \parallel \overline{CF}$

10. $\overline{BE} \parallel \overline{CF}$

11. $\overline{AC} \parallel \overline{DF}$ and $\overline{AE} \parallel \overline{BF}$

12. $x = 120$ (corr. $\angle$);
$x = y + 50$
$120 = y + 50$
$y = 70$

13. $x = 32$; $y = 30$

14. $2x - y = 3x + 2y$
$-1x - 3y = 2$ (Multiply by 3)
$3x - y = 80$

$-3x - 9 = 0$
$\underline{3x - y = 80}$
$-10y = 80$
$y = -8$

$3x + 8 = 80$
$3x = 72$
$x = 24$

15. $2a + 2b + 100 = 180$
$2a + 2b = 80$
$a + b = 40$ ($\div$ by 2)
$(a + b) + x = 180$
$40 + x = 180$
$x = 140$

16. $x^2 - 12 = x(x - 2)$
$x^2 - 12 = x^2 - 2x$
$-12 = -2x$
$x = 6$

17. $x^2 - 3x + 4 + 17x - x^2 - 5 = 111$
$14x - 1 = 222$
$14x = 112$
$x = 8$
$m\angle 3 = 69°$; $m\angle 4 = 67°$; $m\angle 5 = 44°$

18. $3x + y + 5x + 10 = 180$
$3x + y = 5y + 20$
$3x + y = 170$ (Multiply by 4)
$3x - 4y = 20$

$32x + 4y = 680$
$\underline{3x - y = 80}$
$35x = 700$
$x = 20$

$8(20) + y = 170$
$160 + y = 170$
$y = 10$

$m\angle C = 5(10) + 20 = 70°$ $\therefore m\angle B = 110°$

19. S

20. N

21. N

22. S

23. S

24. A

25.

Number of sides	8	12	20	15	10	16	180
Measure of each ext. $\angle$	45	30	18	24	36	22.5	2
Measure of each int. $\angle$	135	150	162	156	144	157.5	178
Number of diagonals	20	54	170	90	35	104	15,930

26.

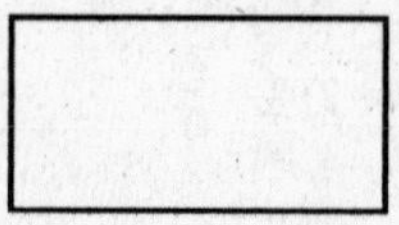

27.

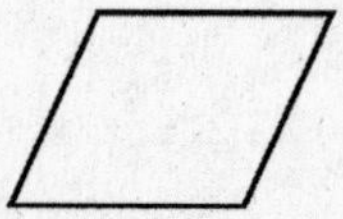

28. Not possible

29.

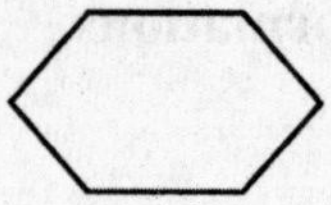

30. Statement: If 2 angles are right angles, then the angles are congruent.
Converse: If 2 angles are congruent, then the angles are right angles.
Inverse: If 2 angles are not right angles, then the angles are not congruent.
Contrapositive: If 2 angles are not congruent, then the angles are not right angles.

31. Statement: If it is not raining, then I am happy.
Converse: If I am happy, then it is not raining.
Inverse: If it is raining, then I am not happy.
Contrapositive: If I am not happy, then it is raining.

32. Contrapositive

33. Given: $\overline{AB} \parallel \overline{CF}$
$\quad\quad\quad \angle 2 \cong \angle 3$
Prove: $\angle 1 \cong \angle 3$

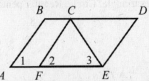

STATEMENTS	REASONS
1. $\overline{AB} \parallel \overline{CF}$	**1.** Given
2. $\angle 1 \cong \angle 2$	**2.** If 2 ∥ lines are cut by a trans., then the corresponding ∠s are ≅.
3. $\angle 2 \cong \angle 3$	**3.** Given
4. $\angle 1 \cong \angle 3$	**4.** Transitive Prop. of Congruence

34. Given: $\angle 1$ is comp. to $\angle 2$
$\quad\quad\quad \angle 2$ is comp. to $\angle 3$
Prove: $\overline{BD} \parallel \overline{AE}$

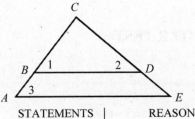

STATEMENTS	REASONS
1. $\angle 1$ is comp to $\angle 2$ $\angle 2$ is comp to $\angle 3$	**1.** Given
2. $\angle 1 \cong \angle 3$	**2.** If 2 ∠s are comp. to the same ∠, then these ∠s are ≅.
3. $\overline{BD} \parallel \overline{AE}$	**3.** If 2 lines are cut by a trans. so that the corresponding ∠s are ≅, then the lines are ∥.

35. Given: $\overline{BE} \perp \overline{DA}$
$\quad\quad\quad \overline{CD} \perp \overline{DA}$
Prove: $\angle 1 \cong \angle 2$

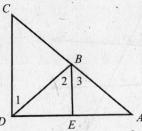

STATEMENTS	REASONS
1. $\overline{BE} \perp \overline{DA}$ $\overline{CD} \perp \overline{DA}$	**1.** Given
2. $\overline{BE} \parallel \overline{CD}$	**2.** If 2 lines are each ⊥ to a 3rd line, then these lines are parallel to each other.
3. $\angle 1 \cong \angle 2$	**3.** If 2 ∥ lines are cut by a trans., then the alternate interior ∠s are ≅.

36. Given: $\angle A \cong \angle C$
$\quad\quad\quad \overline{DC} \parallel \overline{AB}$
Prove: $\overline{DA} \parallel \overline{CB}$

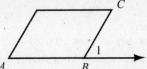

STATEMENTS	REASONS
1. $\angle A \cong \angle C$	**1.** Given
2. $\overline{DC} \parallel \overline{AB}$	**2.** Given
3. $\angle C \cong \angle 1$	**3.** If 2 ∥ lines are cut by a trans., then the alt. int. ∠s are congruent.
4. $\angle A \cong \angle 1$	**4.** Transitive Prop. of Congruence
5. $\overline{DA} \parallel \overline{CB}$	**5.** If 2 lines are cut by a trans. so that corr. ∠s are ≅, then these lines are ∥.

37. Assume $x = -3$.

38. Assume the sides opposite these angles are ≅.

39. Given: $m \nparallel n$

Prove: $\angle 1 \ncong \angle 2$

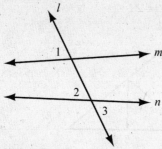

Indirect Proof:

Assume that $\angle 1 \cong \angle 2$. Then $m \parallel n$ since congruent corr. Angles are formed. But this contradicts our hypothesis. Therefore, our assumption must be false and $\angle 1 \ncong \angle 2$.

40. Given: $\angle 1 \ncong \angle 3$

Prove: $m \nparallel n$

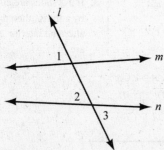

Indirect Proof:

Assume that $m \parallel n$. Then $\angle 1 \cong \angle 3$ since alt. ext. angles are congruent when parallel lines are cut by a transversal. But this contradicts the given fact that $\angle 1 \ncong \angle 3$. Therefore, our assumption must be false and it follows that $m \nparallel n$.

41. Given: $\triangle ABC$

Construct: The line through C parallel to $\overline{AB}$.

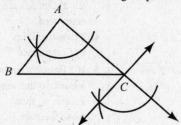

42. Given: $\overline{AB}$

Construct: An equilateral triangle ABC with side $\overline{AB}$.

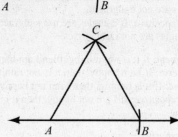

43. a. B, H, W

b. H, S

44. a. Isosceles triangle, Circle, Regular pentagon

b. Circle

45. Congruent

46. a.

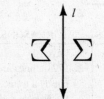

b.

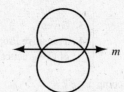

47. 90°

CHAPTER TEST

1. a. $\angle 5$

b. $\angle 3$

2. a. $m\angle 2 + m\angle 8 = 68 + 112 = 180°$
$m\angle 6 + m\angle 9 = 68 + 110 = 178°$
$r \parallel s$

b. $m\angle 2 + m\angle 8 = 68 + 112 = 180°$
$m\angle 6 + m\angle 9 = 68 + 110 = 178°$
$l \nparallel m$

3. not Q

4. $\angle R$ and $\angle S$ are not both right angles.

5. a. If $r \parallel s$ and $s \parallel t$,

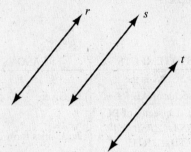

then $r \parallel t$.

b. If $a \perp b$ and $b \perp c$,

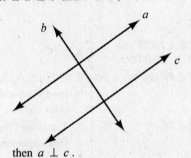

then $a \perp c$.

6.

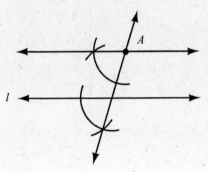

7. a. $65 + 79 + x = 180$
$\quad\quad 144 + x = 180$
$\quad\quad\quad\quad\quad x = 36$

$\quad\quad m\angle B = 36°$

b. $2x + x + 2x + 15 = 180$
$\quad\quad\quad\quad 5x + 15 = 180$
$\quad\quad\quad\quad\quad\quad 5x = 165$
$\quad\quad\quad\quad\quad\quad\; x = 33$

$\quad\quad m\angle B = 33°$

8. a. Pentagon

b. Use $D = \dfrac{n(n-3)}{2}$; 5

9. a. Equiangular hexagon

b. Use $I = \dfrac{180(n-2)}{n}$; 120°

10. A: line; D: line; N: point; O: both; X: both

11. a. Reflection

b. Slide

c. Rotation

12. $\angle 1 \cong \angle 2$ and $\angle 4 \cong \angle 3$
$\quad\; m\angle C = 61°$

13. $x + 28 = 2x - 26$
$\quad\quad\quad\quad x = 54$

14. $m\angle 3 = 80°$ so $m\angle 4 = 50°$

15. $m\angle 3 = 63°$
$\quad\; m\angle 1 = x$
$\quad\; x + 2x + 63 = 180$
$\quad\quad\quad\quad\; 3x = 117$
$\quad\quad\quad\quad\quad\; x = 39$
$\quad\; m\angle 1 = 39°$

16. **1.** Given

$\quad$ **2.** $\angle 2 \cong \angle 3$

$\quad$ **3.** Transitive Prop. of Congruence

$\quad$ **4.** $l \parallel n$

17. Given: $\triangle MNQ$ with $m\angle N = 120°$
$\quad$ Prove: $\angle M$ and $\angle Q$ are not complementary

Indirect Proof:
Assume that $\angle M$ and $\angle Q$ are complementary.
By definition $m\angle M + m\angle Q = 90°$. Also,
$m\angle M + m\angle Q + m\angle N = 180°$ because these are
the three angles of $\triangle MNQ$. By substitution,
$90° + m\angle N = 180°$, so it follows that
$m\angle N = 90°$. But this leads to a contradiction
because it is given that $m\angle N = 120°$. The
assumption must be false, and it follows that
$\angle M$ and $\angle Q$ are not complementary.

18. **1.** Given

$\quad$ **2.** 180°

$\quad$ **3.** $m\angle 1 + m\angle 2 + 90° = 180°$

$\quad$ **4.** 90°

$\quad$ **S5.** $\angle 1$ and $\angle 2$ are complementary.

$\quad$ **R5.** Definition of complementary angles.

19. $m\angle Z = 21°$

Chapter 3: Triangles

SECTION 3.1: Congruent Triangles

1. $\angle A$; $\overline{AB}$; No; No

5. SAS

9. SSS

13. ASA

17. SSS

21. $\overline{AD} \cong \overline{EC}$

25. **a.** Given

 b. $\overline{AC} \cong \overline{AC}$

 c. SSS

29. Given: $\overline{AB} \perp \overline{BC}$ and
 $\overline{AB} \perp \overline{BD}$;
 also $\overline{BC} \cong \overline{BD}$
 Prove: $\triangle ABC \cong \triangle ABD$

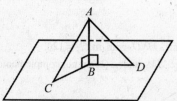

STATEMENTS	REASONS
1. $\overline{AB} \perp \overline{BC}$ and $\overline{AB} \perp \overline{BD}$	1. Given
2. $\angle ABC$ is a right $\angle$ and $\angle ABD$ is a right $\angle$	2. If 2 lines are $\perp$, then they meet to form a right $\angle$.
3. $\angle ABC \cong \angle ABD$	3. Any two right $\angle$s are $\cong$.
4. $\overline{BC} \cong \overline{BD}$	4. Given
5. $\overline{AB} \cong \overline{AB}$	5. Identity
6. $\triangle ABC \cong \triangle ABD$	6. SAS

33. Yes; SAS or SSS

37. **a.** $\triangle CBE$, $\triangle ADE$, $\triangle CDE$

 b. $\triangle ADC$

 c. $\triangle CBD$

41.

STATEMENTS	REASONS
$\angle ABC$; $\overline{RS}$ is the $\perp$	
1. bisector of $\overline{AB}$; $\overline{RT}$ is the $\perp$ bisector of $\overline{BC}$	1. Given
2. $\angle RSA \cong \angle RSB$ and $\angle RTB \cong \angle RTC$	2. $\perp$ lines form $\cong$ adj. $\angle$s
3. $\overline{AS} \cong \overline{BS}$; $\overline{BT} \cong \overline{TC}$	3. Bisecting a seg. forms 2 $\cong$ segments.
4. $\overline{RS} \cong \overline{RS}$ and $\overline{RT} \cong \overline{RT}$	4. Identity
5. $\triangle RSA \cong \triangle RSB$ and $\triangle RTB \cong \triangle RTC$	5. SAS
6. $\overline{AR} \cong \overline{BR}$ and $\overline{BR} \cong \overline{RC}$	6. If 2 triangles are $\cong$, the corresponding sides are $\cong$.
7. $\overline{AR} \cong \overline{RC}$	7. Transitive Property for Congruence

SECTION 3.2: Corresponding Parts of Congruent Triangles

1. Given: ∠1 and ∠2 are right ∠s.
$\overline{CA} \cong \overline{DA}$

Prove: △$ABC \cong$ △ABD

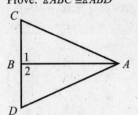

STATEMENTS	REASONS
1. ∠1 and ∠2 are right ∠s and $\overline{CA} \cong \overline{DA}$	1. Given
2. $\overline{AB} \cong \overline{AB}$	2. Identity
3. △$ABC \cong$△ABD	3. HL

5. Given: ∠R and ∠V are right ∠s.
∠1 ≅ ∠2

Prove: △$RST \cong$ △VST

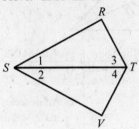

STATEMENTS	REASONS
1. ∠R and ∠V are right ∠s and ∠1 ≅ ∠2	1. Given
2. ∠R ≅ ∠V	2. All right ∠s are ≅.
3. $\overline{ST} \cong \overline{ST}$	3. Identity
4. △$RST \cong$△VST	4. AAS

9. m∠2 = 48°; m∠3 = 48°
m∠5 = 42°; m∠6 = 42°

13. Given: ∠s P and R are rt. ∠s
M is the midpoint of $\overline{PR}$

Prove: ∠N ≅ ∠Q

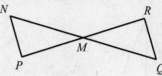

STATEMENTS	REASONS
1. ∠s P and R are rt. ∠s	1. Given
2. ∠P ≅ ∠R	2. All rt. ∠s are ≅.
3. M is the midpoint of $\overline{PR}$	3. Given
4. $\overline{PM} \cong \overline{MR}$	4. Midpoint of a segment forms 2 ≅ segments.
5. ∠NMP ≅ ∠QMR	5. If 2 lines intersect, the vertical angles formed are ≅.
6. △$NPM \cong$△QRM	6. ASA
7. ∠N ≅ ∠Q	7. CPCTC

17. $c = 5$

21. $c = \sqrt{41}$

25. Given: E is the midpoint of $\overline{FG}$;
$\overline{DF} \cong \overline{DG}$

Prove: $\overline{DE} \perp \overline{FG}$

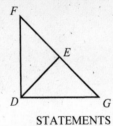

STATEMENTS	REASONS
1. E is the midpoint of $\overline{FG}$	1. Given
2. $\overline{FE} \cong \overline{EG}$	2. The midpoint of a segment forms 2 ≅ segments.
3. $\overline{DF} \cong \overline{DG}$	3. Given
4. $\overline{DE} \cong \overline{DE}$	4. Reflexive
5. △$FDE \cong$△GDE	5. SSS
6. ∠DEF ≅ ∠DEG	6. CPCTC
7. $\overline{DE} \perp \overline{FG}$	7. If 2 lines meet to form ≅ adj. ∠s, then the lines are ⊥.

29. Given: $\overrightarrow{RW}$ bisects $\angle SRU$;
      also $\overline{RS} \cong \overline{RU}$

Prove: $\triangle TRU \cong \triangle VRS$

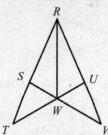

STATEMENTS	REASONS
1. $\overrightarrow{RW}$ bisects $\angle RSU$	1. Given
2. $\angle SRW \cong \angle URW$	2. If a ray bisects an $\angle$ then 2 $\cong$ $\angle$s are formed.
3. $\overline{RS} \cong \overline{RU}$	3. Given
4. $\overline{RW} \cong \overline{RW}$	4. Identity
5. $\triangle RSW \cong \triangle RUW$	5. SAS
6. $\angle RSW \cong \angle RUW$	6. CPCTC
7. $\angle TRV \cong \angle VRS$	7. Identity
8. $\triangle TRU \cong \triangle VRS$	8. ASA

33. 751 feet

SECTION 3.3: Isosceles Triangles

1. Isosceles

5. $m\angle U = 69°$

9. $L = E$ (equivalent)

13. Underdetermined

17. Determined

21. $m\angle 2 = 68°$ (Base $\angle$s in an isosceles $\triangle$ are $\cong$)
      Also, $m\angle 1 + m\angle 2 + m\angle 3 = 180$
      $m\angle 1 + 68 + 68 = 180$
      $m\angle 1 + 136 = 180$
      $m\angle 1 = 44°$

25. Let the measure of the vertex angle be x. Then the measure of the base angles are each $x + 12$.
$$x + (x + 12) + (x + 12) = 180$$
$$3x + 24 = 180$$
$$3x = 156$$
$$x = 52$$

$m\angle A = 52°$; $m\angle B = 64°$; $m\angle C = 64°$

29. · 12

33. 1. Given

   2. $\angle 3 \cong \angle 2$

   3. $\angle 1 \cong \angle 2$

   4. If 2 $\angle$s of one $\triangle$ are $\cong$, then the opposite sides are $\cong$.

37. Given: Isosceles $\triangle MNP$ with vertex P;
      Isosceles $\triangle MNQ$ with vertex Q

Prove: $\triangle MQP \cong \triangle NQP$

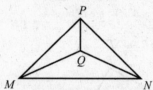

STATEMENTS	REASONS
1. Isosceles $\triangle MNP$ with vertex P	1. Given
2. $\overline{MP} \cong \overline{NP}$	2. An isosceles $\triangle$ has 2 $\cong$ sides.
3. Isosceles $\triangle MNQ$ with vertex Q	3. Given
4. $\overline{MQ} \cong \overline{NQ}$	4. Same as (2)
5. $\overline{PQ} \cong \overline{PQ}$	5. Identity
6. $\triangle MQP \cong \triangle NQP$	6. SSS

41. 75°

SECTION 3.4: Basic Constructions
Justified

1.

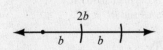

5.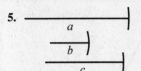

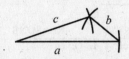

9.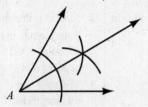

13.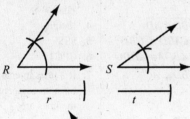

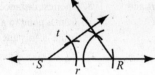

17.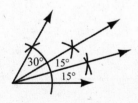

21.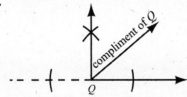

25.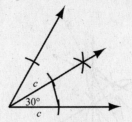

29. Given: Line *m*, with point *P* on *m*

$\overline{PQ} \cong \overline{PR}$ (by construction)

$\overline{QS} \cong \overline{RS}$ (by construction)

Prove: $\overrightarrow{SP} \perp m$

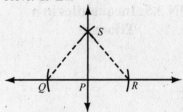

STATEMENTS	REASONS
1. Line *m* with point *P* on line *m* $\overline{PQ} \cong \overline{PR}$ and $\overline{QS} \cong \overline{RS}$	1. Given
2. $\overline{SP} \cong \overline{SP}$	2. Identity
3. $\triangle SPQ \cong \triangle SPR$	3. SSS
4. $\angle SPQ \cong \angle SPR$	4. CPCTC
5. $\overrightarrow{SP} \perp m$	5. If 2 lines intersect to form $\cong$ adjacent $\angle$s, then the lines are $\perp$.

33. 150°

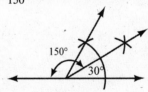

37. Yes

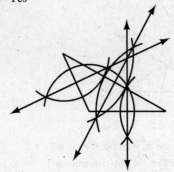

SECTION 3.5: Inequalities in a Triangle

1. False

5. True

9. True

13. **a.** Possible

    b. Not possible $(8 + 9 = 17)$

    c. Not possible $(8 + 9 < 18)$

17. Isosceles Obtuse triangle $(m\angle Z = 100°)$

21. Largest $\angle$ is $72°$ (two of these); smallest is $36°$

25. **1.** $m\angle ABC > m\angle DBE$ and $m\angle CBD > m\angle EBF$

    3. Angle-Addition Postulate

    4. $m\angle ABD > m\angle DBF$

29. $BC < EF$

33. $x + 2 < y < 5x + 12$

37. The length of the median from the vertex of an isosceles triangle is less than the length of either of the legs.

Given: $\triangle ABC$ is isosceles with $\overline{AB} \cong \overline{BC}$

          $\overline{BD}$ is the median to $\overline{AC}$

Prove: $BD < AB$ and $BD < BC$

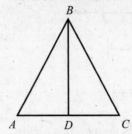

STATEMENTS	REASONS
1. $\triangle ABC$ is isosceles with $\overline{AB} \cong \overline{BC}$	**1.** Given
2. D is the midpoint of $\overline{AC}$	**2.** Median is drawn from a vertex to the midpoint of the opposite side.
3. $\overline{AD} \cong \overline{DC}$	**3.** Midpoint of a segment forms 2 $\cong$ segments.
4. $\overline{BD} \cong \overline{BD}$	**4.** Identity
5. $\triangle ABD \cong \triangle CBD$	**5.** SSS
6. $\angle BDA \cong \angle BDC$	**6.** CPCTC
7. $\overline{BD} \perp \overline{AC}$	**7.** If 2 lines intersect to form $\cong$ adjacent $\angle$s, the lines are $\perp$.
8. $BD < AB$ and $BD < BC$	**8.** Shortest distance from a point to a line is the $\perp$ distance.

CHAPTER REVIEW

1. Given: $\angle AEB \cong \angle DEC$
$\overline{AE} \cong \overline{ED}$
Prove: $\triangle AEB \cong \triangle DEC$

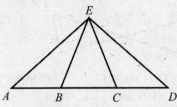

STATEMENTS	REASONS
1. $\angle AEB \cong \angle DEC$	1. Given
2. $\overline{AE} \cong \overline{ED}$	2. Given
3. $\angle A \cong \angle D$	3. If 2 sides of a $\triangle$ are $\cong$, then the $\angle$s opposite these sides are also $\cong$.
4. $\triangle AEB \cong \triangle DEC$	4. ASA

2. Given: $\overline{AB} \cong \overline{EF}$
$\overline{AC} \cong \overline{DF}$
$\angle 1 \cong \angle 2$
Prove: $\angle B \cong \angle E$

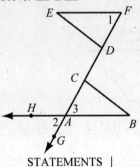

STATEMENTS	REASONS
1. $\overline{AB} \cong \overline{EF}$	1. Given
2. $\overline{AC} \cong \overline{DF}$; $\angle 1 \cong \angle 2$	2. Given
3. $\angle 2 \cong \angle 3$	3. If 2 lines intersect, then the vertical $\angle$s formed are $\cong$.
4. $\angle 1 \cong \angle 3$	4. Transitive Prop. for $\cong$.
5. $\triangle ABC \cong \triangle FED$	5. SAS
6. $\angle B \cong \angle E$	6. CPCTC

3. Given: $\overline{AD}$ bisects $\overline{BC}$
$\overline{AB} \perp \overline{BC}$
$\overline{DC} \perp \overline{BC}$
Prove: $\overline{AE} \cong \overline{ED}$

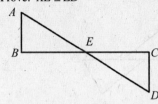

STATEMENTS	REASONS
1. $\overline{AB} \cong \overline{EF}$	1. Given
2. $\overline{AC} \cong \overline{DF}$; $\angle 1 \cong \angle 2$	2. Given
3. $\angle 2 \cong \angle 3$	3. If 2 lines intersect, then the vertical $\angle$s formed are $\cong$.
4. $\angle 1 \cong \angle 3$	4. Transitive Prop. for $\cong$.
5. $\triangle ABC \cong \triangle FED$	5. SAS
6. $\angle B \cong \angle E$	6. CPCTC

4. Given: $\overline{OA} \cong \overline{OB}$
$\overline{OC}$ is the median to $\overline{AB}$
Prove: $\overline{OC} \perp \overline{AB}$

STATEMENTS	REASONS
1. $\overline{OA} \cong \overline{OB}$	1. Given
2. $\overline{OC}$ is the median to $\overline{AB}$	2. Given
3. C is the midpoint of $\overline{AB}$	3. The median of a $\triangle$ is a segment drawn from a vertex to the midpoint of the opp. side.
4. $\overline{AC} \cong \overline{CB}$	4. Midpoint of seg. form 2 $\cong$ segments.
5. $\overline{OC} \cong \overline{OC}$	5. Identity
6. $\triangle AOC \cong \triangle BOC$	6. SSS
7. $\angle OCA \cong \angle OCB$	7. CPCTC
8. $\overline{OC} \perp \overline{AB}$	8. If 2 lines meet to form $\cong$ adj. $\angle$s, then the lines are $\perp$.

5. Given: $\overline{AB} \cong \overline{DE}$
$\overline{AB} \parallel \overline{DE}$
$\overline{AC} \cong \overline{DF}$
Prove: $\overline{BC} \parallel \overline{FE}$

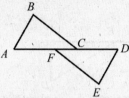

STATEMENTS	REASONS
1. $\overline{AB} \cong \overline{DE}$ and $\overline{AB} \parallel \overline{DE}$	**1.** Given
2. $\angle A \cong \angle D$	**2.** If 2 $\parallel$ lines are cut by a trans., then the alt. int. $\angle$s are $\cong$.
3. $\overline{AC} \cong \overline{DF}$	**3.** Given
4. $\triangle BAC \cong \triangle EDF$	**4.** SAS
5. $\angle BCA \cong \angle EFD$	**5.** CPCTC
6. $\overline{BC} \parallel \overline{FE}$	**6.** If 2 lines are cut by a trans. so that alt. int. $\angle$s are $\cong$, then the lines are $\parallel$.

6. Given: *B* is the midpoint of $\overline{AC}$
$\overline{BD} \perp \overline{AC}$
Prove: $\triangle ADC$ is isosceles

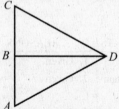

STATEMENTS	REASONS
1. *B* is the midpoint of $\overline{AC}$	**1.** Given
2. $\overline{CB} \cong \overline{BA}$	**2.** Midpoint of a segment forms 2 $\cong$ segments.
3. $\overline{BD} \perp \overline{AC}$	**3.** Given
4. $\angle DBC \cong \angle DBA$	**4.** If 2 lines are $\perp$, they meet to form $\cong$ adj. $\angle$s.
5. $\overline{BD} \cong \overline{BD}$	**5.** Identity
6. $\triangle CBD \cong \triangle ABD$	**6.** SAS
7. $\overline{DC} \cong \overline{DA}$	**7.** CPCTC
8. $\triangle ADC$ is isosceles	**8.** If a $\triangle$ has 2 $\cong$ sides, it is an isos. $\triangle$.

7. Given: $\overline{JM} \perp \overline{GM}$
$\overline{GK} \perp \overline{KJ}$
$\overline{GH} \cong \overline{HJ}$
Prove: $\overline{GM} \cong \overline{JK}$

STATEMENTS	REASONS
1. $\overline{JM} \perp \overline{GM}$ and $\overline{GK} \perp \overline{KJ}$	**1.** Given
2. $\angle M$ is a rt. $\angle$ and $\angle K$ is a rt. $\angle$	**2.** If 2 lines are $\perp$, they meet to form a rt. $\angle$.
3. $\angle M \cong \angle K$	**3.** Any 2 rt. $\angle$s are $\cong$.
4. $\overline{GH} \cong \overline{HJ}$	**4.** Given
5. $\angle GHM \cong \angle JHK$	**5.** If 2 lines intersect, the vertical $\angle$s formed are $\cong$.
6. $\triangle GHM \cong \triangle JHK$	**6.** AAS
7. $\overline{GM} \cong \overline{JK}$	**7.** CPCTC

8. Given: $\overline{TN} \cong \overline{TR}$
$\overline{TO} \perp \overline{NP}$
$\overline{TS} \perp \overline{PR}$
$\overline{TO} \cong \overline{TS}$
Prove: $\angle N \cong \angle R$

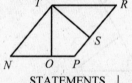

STATEMENTS	REASONS
1. $\overline{TN} \cong \overline{TR}$	**1.** Given
2. $\overline{TO} \perp \overline{NP}$; $\overline{TS} \perp \overline{PR}$	**2.** Given
3. $\angle TON$ is a rt. $\angle$ and $\angle TSR$ is a rt. $\angle$.	**3.** If 2 lines are $\perp$, they meet to form a rt. $\angle$.
4. $\overline{TO} \cong \overline{TS}$	**4.** Given
5. $\triangle TON \cong \triangle TSR$	**5.** HL
6. $\angle N \cong \angle R$	**6.** CPCTC

9. Given: $\overline{YZ}$ is the base of an isosceles triangle
   $\overline{XA} \parallel \overline{YZ}$

   Prove: $\angle 1 \cong \angle 2$

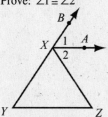

STATEMENTS	REASONS
1. $\overline{YZ}$ is the base of an isosceles △	1. Given
2. $\angle Y \cong \angle Z$	2. Base ∠s of an isos. △ are ≅.
3. $\overline{XA} \parallel \overline{YZ}$	3. Given
4. $\angle 1 \cong \angle Y$	4. If 2 ∥ lines are cut by a trans., then the corresp. ∠s are ≅.
5. $\angle 2 \cong \angle Z$	5. If 2 ∥ lines are cut by a trans., then the corresp. ∠s are ≅.
6. $\angle 1 \cong \angle 2$	6. Transitive Prop. for ≅.

10. Given: $\overline{AB} \parallel \overline{DC}$
    $\overline{AB} \cong \overline{DC}$
    C is the midpoint of $\overline{BE}$

    Prove: $\overline{AC} \parallel \overline{DE}$

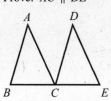

STATEMENTS	REASONS
1. $\overline{AB} \cong \overline{DE}$ and $\overline{AB} \parallel \overline{DC}$	1. Given
2. $\angle B \cong \angle DCE$	2. If 2 ∥ lines are cut by a trans., then the corresp. ∠s are ≅.
3. $\overline{AB} \cong \overline{DC}$	3. Given
4. C is the midpoint of $\overline{BE}$	4. Given
5. $\overline{BC} \cong \overline{CE}$	5. Midpoint of a seg. forms 2 ≅ segments.
6. $\triangle ABC \cong \triangle DCE$	6. SAS
7. $\angle ACB \cong \angle E$	7. CPCTC
8. $\overline{AC} \parallel \overline{DE}$	8. If 2 lines are cut by a trans. so that the corresp. ∠s are ≅, then the lines are ∥.

11. Given: $\angle BAD \cong \angle CDA$
    $\overline{AB} \cong \overline{CD}$

    Prove: $\overline{AE} \cong \overline{ED}$

STATEMENTS	REASONS
1. $\angle BAD \cong \angle CDA$	1. Given
2. $\overline{AB} \cong \overline{CD}$	2. Given
3. $\overline{AD} \cong \overline{AD}$	3. Identity
4. $\triangle BAD \cong \triangle CDA$	4. SAS
5. $\angle 1 \cong \angle 2$	5. CPCTC
6. $\overline{AE} \cong \overline{ED}$	6. If 2 ∠s of a triangle are ≅, then the sides opp. these ∠s are also ≅.

12. Given: $\overline{BE}$ is altitude to $\overline{AC}$
    $\overline{AD}$ is altitude to $\overline{CE}$
    $\overline{BC} \cong \overline{CD}$

    Prove: $\overline{BE} \cong \overline{AD}$

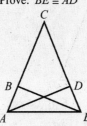

STATEMENTS	REASONS
1. $\overline{BE}$ is altitude to $\overline{AC}$ $\overline{AD}$ is altitude to $\overline{CE}$	1. Given
2. $\overline{BE} \perp \overline{AC}$ and $\overline{AD} \perp \overline{CE}$	2. An altitude is a line segment drawn from a vertex ⊥ to opp. side.
3. $\angle CBE$ is a rt. ∠ and $\angle CDA$ is a rt. ∠	3. If 2 lines are ⊥, they meet to form a rt. ∠.
4. $\angle CBE \cong \angle CDA$	4. Any 2 rt. ∠s are ≅.
5. $\overline{BC} \cong \overline{CD}$	5. Given
6. $\angle C \cong \angle C$	6. Identity
7. $\triangle CBE \cong \triangle CDA$	7. ASA
8. $\overline{BE} \cong \overline{AD}$	8. CPCTC

13. Given: $\overline{AB} \cong \overline{CD}$

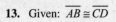

 $\angle BAD \cong \angle CDA$

Prove: $\triangle AED$ is isosceles

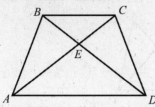

STATEMENTS	REASONS
1. $\overline{AB} \cong \overline{CD}$	1. Given
2. $\angle BAD \cong \angle CDA$	2. Given
3. $\overline{AD} \cong \overline{AD}$	3. Identity
4. $\triangle BAD \cong \triangle CDA$	4. SAS
5. $\angle CAD \cong \angle BDA$	5. CPCTC
6. $\overline{AE} \cong \overline{ED}$	6. If 2 $\angle$s of a $\triangle$ are $\cong$, then the sides opp. these $\angle$s are also $\cong$.
7. $\triangle AED$ is isosceles	7. If a $\triangle$ has 2 $\cong$ sides, it is an isosceles $\triangle$.

14. Given: $\overrightarrow{AC}$ bisects $\angle BAD$

Prove: $AD > CD$

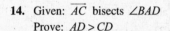
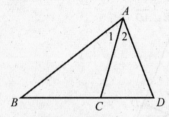

STATEMENTS	REASONS
1. $\overrightarrow{AC}$ bisects $\angle BAD$	1. Given
2. $m\angle 1 \cong m\angle 2$	2. If a ray bisects an $\angle$, it forms 2 $\angle$s of = measure.
3. $m\angle ACD > m\angle 1$	3. The measure of an ext. $\angle$ of a $\triangle$ is greater than the measure of either of the nonadjacent interior angles.
4. $m\angle ACD > m\angle 2$	4. Substitution
5. $AD > CD$	5. If the measure of one angle of a $\triangle$ is greater than the measure of a second angle, then the side which is opposite the 1st angle is longer than the side which is opp. the second angle.

15. a. $\overline{PR}$

  b. $\overline{PQ}$

16. $\overline{BC}$, $\overline{AC}$, $\overline{AB}$

17. $\angle R$, $\angle Q$, $\angle P$

18. $\overline{AD}$

19. (b)

20. 5 and 35

21. $20°$

22. $115°$

23. $3x + 10 = \dfrac{5}{2}x + 18$

$\qquad \dfrac{1}{2}x = 8$

$\qquad\quad x = 16$

$\qquad m\angle 4 = \dfrac{5}{2}(16) + 18 = 58°$

$\qquad m\angle C = 64°$

24. $10 + x + 6 + 2x - 3 = 40$

$\qquad\qquad 3x + 13 = 40$

$\qquad\qquad\quad 3x = 27$

$\qquad\qquad\quad\ x = 9$

$AB = 10$; $BC = 15$; $AC = 15$: the triangle is isosceles.

25. Either $AB = BC$ or $AB = AC$ or $BC = AC$.

If $AB = BC$, $\quad y + 7 = 3y + 5$

$\qquad\qquad\qquad -2y = -2$

$\qquad\qquad\qquad\quad y = 1$

If $AB = AC$, $\quad y + 7 = 9 - y$

$\qquad\qquad\qquad\ 2y = 2$

$\qquad\qquad\qquad\ \ y = 1$

If $BC = AC$, $\quad 3y + 5 = 9 - y$

$\qquad\qquad\qquad\ 4y = 4$

$\qquad\qquad\qquad\ \ y = 1$

If $y = 1$, $AB = 8$; $BC = 8$; $AC = 8$; the triangle is also equilateral.

26. If $m\angle 1 = 5x$, then the $m\angle 2 = 180 - 5x$

$\quad m\angle 4 = m\angle 2 = 180 - 5x$.

But $m\angle 3 = m\angle 4$, therefore

$2x + 12 = 180 - 5x$

$\qquad 7x = 168$

$\qquad\ x = 24$

$m\angle 2 = 180 - 5(24) = 60°$

27. Construct an angle that measures 75°.

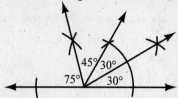

28. Construct a right triangle that has acute angle A and hypotenuse of length c.

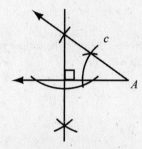

29. Construct another isosceles triangle in which the base angles are half as large as the given base angles.

CHAPTER TEST

1. a. Since $\triangle ABC \cong \triangle DEF$, then $\text{m}\angle A \cong \text{m}\angle D$, $\text{m}\angle B \cong \text{m}\angle E$ and $\text{m}\angle C \cong \text{m}\angle F$.

$\text{m}\angle A \cong 37°$ and $\text{m}\angle E \cong 68°$

$\text{m}\angle F = 180 - 37 - 68 = 75°$

b. Since $\triangle ABC \cong \triangle DEF$, then $\overline{AB} \cong \overline{DE}$, $\overline{BC} \cong \overline{EF}$, and $\overline{AC} \cong \overline{DF}$.

$AB = 7.3$, $BC = 4.7$, and $AC = 6.3$

$EF = 4.7$ cm

2. a. $\overline{XY}$

b. $\angle Y$

3. a. SAS

b. ASA

4. Corresponding parts of congruent triangles are congruent.

5. a. No

b. Yes

6. Yes

7. a. $c = 10$

b. $b = \sqrt{c^2 - a^2}$
$b = \sqrt{8^2 - 6^2}$
$b = \sqrt{64 - 36}$
$b = \sqrt{28} = 2\sqrt{7}$

8. a. $\overline{AM} \cong \overline{BM}$

b. No

9. a. $\text{m}\angle V = 180 - 71 - 71 = 38°$

b. $7x + 2 = 9(x - 2)$
$7x + 2 = 9x - 18$
$20 = 2x$
$x = 10$
$\text{m}\angle T = 7(10) + 2 = 72°$
$\text{m}\angle U = 9(10 - 2) = 72°$
$\text{m}\angle V = 180 - 72 - 72 = 36°$

10. a. $VU = 7.6$ in.

b. $4x + 1 = 6x - 10$
$11 = 2x$
$x = \dfrac{11}{2}$
$VT = 4\left(\dfrac{11}{2}\right) + 1 = 23$
$TU = 2\left(\dfrac{11}{2}\right) = 11$
$VU = 6\left(\dfrac{11}{2}\right) - 10 = 23$
$P = 23 + 11 + 23 = 57$

11. a. Construct an angle that measures 60°.

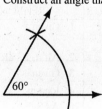

b. Construct an angle that measures 60°.

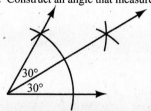

12.

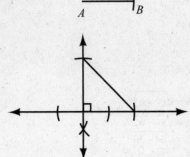

13. a. $\overline{BC}$

    b. $\overline{CA}$

14. $m\angle V > m\angle U > m\angle T$

15. $EB = \sqrt{(AE)^2 + (AB)^2}$

$EB = \sqrt{(4+3)^2 + 5^2}$

$EB = \sqrt{49 + 25} = \sqrt{74}$

$DC = \sqrt{(AD)^2 + (AC)^2}$

$DC = \sqrt{4^2 + (5+2)^2}$

$DC = \sqrt{16 + 49} = \sqrt{65}$

$EG > DC$ since $EB = \sqrt{74}$ and $DC = \sqrt{65}$.

16.

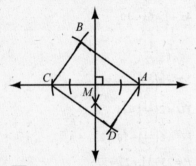

$\overline{DA}$

17.

STATEMENTS	REASONS
1. $\angle R$ and $\angle V$ are rt. $\angle$s	1. Given
2. $\angle R \cong \angle V$	2. All rt. $\angle$s are $\cong$
3. $\angle 1 \cong \angle 2$	3. Given
4. $\overline{ST} \cong \overline{ST}$	4. Identity
5. $\triangle RST \cong \triangle VST$	5. AAS

18. R1. Given

    R2. If 2 $\angle$s of a $\triangle$ are $\cong$, the opposite sides are $\cong$.

    S3. $\angle 1 \cong \angle 3$

    R4. ASA

    S5. $\overline{US} \cong \overline{UT}$

    S6. $\triangle STU$ is an isosceles triangle

19. Let each leg = a, and half the base = b.

Then $2a + 2b = 32$ or $a + b = 16$ or $a = 16 \cdot b$.

But $\sqrt{8^2 + b^2} = a$ so

$\sqrt{8^2 + b^2} = 16 \cdot b$ Squaring both sides gives
$8^2 + b^2 = 256 \cdot 32b + b^2$ or

    $64 \cdot 256 = \cdot 32b$

    $\cdot 192 = \cdot 32b$

    $b = 6 \therefore a = 10$

Chapter 4: Quadrilaterals

SECTION 4.1: Properties of a Parallelogram

1. a. $AB = DC$

   b. $m\angle A = m\angle C$

5. In parallelogram $ABCD$, $AB = DC$.
Therefore,
$$3x + 2 = 5x - 2$$
$$4 = 2x$$
$$x = 2$$
$$AB = DC = 8$$
$$BC = AD = 9$$

11. $\overline{AC}$

15. True

19. The resulting quadrilateral appears to be a parallelogram.

23. **1.** Given

   2. $\overline{RV} \perp \overline{VT}$ and $\overline{ST} \perp \overline{VT}$

   3. $\overline{RV} \parallel \overline{ST}$

   4. $RSTV$ is a parallelogram

27. The opposite angles of a parallelogram are congruent.
Given: Parallelogram $ABCD$
Prove: $\angle BAD \cong \angle BCD$ and
$\angle ABC \cong \angle ADC$

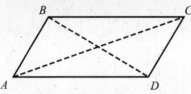

STATEMENTS	REASONS
1. Parallelogram $ABCD$	**1.** Given
2. Draw diagonal $\overline{BD}$	**2.** Through 2 points there is exactly one line.
3. $\triangle ABD \cong \triangle CDB$	**3.** A diagonal of a parallelogram separates it into $2 \cong \triangle s$.
4. $\angle BAD \cong \angle BCD$	**4.** CPCTC
5. Draw in diagonal $\overline{AC}$	**5.** Same as (2)
6. $\triangle ABC \cong \triangle CDA$	**6.** Same as (3)
7. $= m\angle A + m\angle B$	**7.** CPCTC

31. $\angle P$ is a right angle.

35. 255 mph

39. $x + 16 + 2(x + 1) + \dfrac{3x}{2} - 11 + \dfrac{7x}{3} - 16 = 360$

$x + 2x + 2 + \dfrac{3x}{2} - 11 + \dfrac{7x}{3} = 360$

$3x + \dfrac{3x}{2} + \dfrac{7x}{3} = 369$

Multiply by 6, the LCD

$18x + 9x + 14x = 2214$

$41x = 2214$

$x = 54$; $m\angle A = 70°$; $m\angle B = 110°$; $m\angle C = 70°$; $m\angle D = 110°$; $ABCD$ is a parallelogram

SECTION 4.2: The Parallelogram and Kite

1. a. Yes

b. No

5. a. Kite

b. Parallelogram

9. 6.18

13. 10

17. Parallel and congruent

21. Given: *M-Q-T* and *P-Q-R* so that
MNPQ and *QRST* are parallelograms
Prove: $\angle N \cong \angle S$

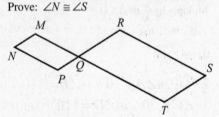

STATEMENTS	REASONS
1. *M-Q-T* and *P-Q-R* so that *MNPQ* and *QRST* are parallelograms.	1. Given
2. $\angle N \cong \angle MQP$	2. Opposite $\angle$s in a parallelogram are $\cong$.
3. $\angle MPQ \cong \angle RQT$	3. If 2 lines intersect, the vertical $\angle$s formed are $\cong$.
4. $\angle RQT \cong \angle S$	4. Same as (2)
5. $\angle N \cong \angle S$	5. Transitive Prop. for $\cong$.

25. If both pairs of opposite sides of a quadrilateral are congruent, then the quadrilateral is a parallelogram.
Given: Quad. *ABCD* with
$\overline{AB} \cong \overline{CD}$ and $\overline{BC} \cong \overline{AD}$
Prove: *ABCD* is a parallelogram

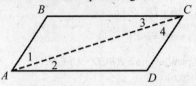

STATEMENTS	REASONS
1. Quad *ABCD* with $\overline{AB} \cong \overline{CD}$ and $\overline{BC} \cong \overline{AD}$	1. Given
2. Draw in $\overline{AC}$	2. Through 2 points there is exactly one line.
3. $\overline{AC} \cong \overline{AC}$	3. Identity
4. $\triangle ABC \cong \triangle CDA$	4. SSS
5. $\angle 1 \cong \angle 4$ and $\angle 2 \cong \angle 3$	5. CPCTC
6. $\overline{AB} \parallel \overline{CD}$ and $\overline{BC} \parallel \overline{AD}$	6. If 2 lines are cut by a trans. so that alt. int. $\angle$s are $\cong$, then the lines are $\parallel$.
7. *ABCD* is a parallelogram	7. If a quad. has both pairs of opposite sides $\parallel$, the quad. is a parallelogram.

29.
$$MN = \frac{1}{2} \cdot ST$$
$$2y - 3 = \frac{1}{2} \cdot 3y$$
$$2y - 3 = \frac{3}{2} y$$
$$\frac{1}{2} y = 3$$
$$y = 6$$
$$MN = 9$$
$$ST = 18$$

35. $m\angle B = 360 - m\angle A - m\angle C - m\angle D$
$= 360 - 30 - 30 - 30$
$= 270°$

SECTION 4.3: The Rectangle, Square and Rhombus

1. $m\angle A = 60°$; $m\angle ABC = 120°$

5. The quadrilateral is a rhombus.

9. $2x + 7 = 3x + 2$
  $x = 5$
  $AD = BC = 3(5) + 4 = 19$

13. $QP = \sqrt{72} = 6\sqrt{2}$
  $MN = \sqrt{72} = 6\sqrt{2}$

17. $AD = \sqrt{34}$

21. True

25. (a)

29. A rectangle is a parallelogram. Therefore, the opposite angles are congruent and the consecutive angles are supplementary. If a rectangle has one right angle which measures 90°, the other three angles must also be 90°. Therefore, all the angles in a rectangle are right angles.

33. A diagonal of a rhombus bisects tow angles of the rhombus.
Given: *ABCD* is a rhombus
Prove: $\overrightarrow{AC}$ bisects $\angle BAD$
    and $\overrightarrow{CA}$ bisects $\angle BCD$

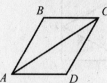

STATEMENTS	REASONS
1. *ABCD* is a rhombus	1. Given
2. $\overline{AB} \cong \overline{BC} \cong \overline{CD} \cong \overline{AD}$	2. All sides of a rhombus are $\cong$.
3. *ABCD* is a parallelogram	3. A rhombus is a parallelogram with 2 $\cong$ adj. sides.
4. $\angle B \cong \angle D$	4. Opposite angles of a parallelogram are $\cong$.
5. $\triangle ABC \cong \triangle ADC$	5. SAS
6. $\angle BAC \cong \angle DAC$ and $\angle BCA \cong \angle DCA$	6. CPCTC
7. $\overrightarrow{AC}$ bisects $\angle BAD$ and $\overrightarrow{CA}$ bisects $\angle BCD$	7. If a ray divides an $\angle$ into 2 $\cong$ $\angle$s, then the ray bisects the $\angle$.

37. If the midpoints of the sides of a rectangle are joined in order, then the quadrilateral formed is a rhombus.

Given: Rect. *ABCD* with *M*, *N*, *O*, and *P* the midpoints of the sides.

Prove: *MNOP* is a rhombus

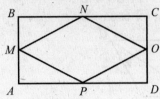

STATEMENTS	REASONS
1. Rect. *ABCD* with *M*, *N*, *O* and *P* the midpoints of the sides.	1. Given
2. *MNOP* is a parallelogram	2. From Exercise 36 of Section 4.2, when the midpoints of the consecutive sides of a quadrilateral are joined in order, the resulting quadrilateral is a parallelogram.
3. ∠s *A* and *B* are rt. ∠s	3. All angles of a rect. are rt. ∠s.
4. ∠*A* ≅ ∠*B*	4. Any 2 right ∠s are ≅.
5. $\overline{MB} \cong \overline{MA}$	5. The midpoint of a segment forms 2 ≅ segments.
6. $\overline{BC} \cong \overline{AD}$	6. Opposite sides of a parallelogram are ≅.
7. $\overline{BN} \cong \overline{AP}$	7. If two segments are ≅, then their midpoints separate these segments into four ≅ segments.
8. △*MBN* ≅ △*MAP*	8. SAS
9. $\overline{MN} \cong \overline{MP}$	9. CPCTC
10. *MNOP* is rhombus	10. If a parallelogram has 2 adj. sides ≅, then the parallelogram is a rhombus.

41. rhombus

SECTION 4.4: The Trapezoid

1. $m\angle D = 180 - 58 = 122°$

$m\angle B = 180 - 125 = 55°$

5. The quadrilateral is a rhombus.

9. a. Yes

    b. No

13. $MN = \frac{1}{2}(AB + DC)$

$9.5 = \frac{1}{2}(8.2 + DC)$

$19 = 8.2 + DC$

$DC = 10.8$

17. Given: *ABCD* is an isosceles trapezoid.

Prove: △*ABE* is isosceles

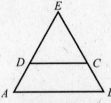

STATEMENTS	REASONS
1. *ABCD* is an isosceles trap.	1. Given
2. ∠*A* ≅ ∠*B*	2. Lower base angles of an isosceles trap. are ≅.
3. $\overline{EB} \cong \overline{EA}$	3. If 2 ∠s of a △ are ≅, then the sides opposite these ∠s are also ≅.
4. △*ABE* is isosceles	4. If a △ has 2 ≅ sides, it is an isosceles △.

21. $(QP)^2 = (MQ)^2 + (MP)^2$
$13^2 = 5^2 + (MP)^2$
$169 = 25 + (MP)^2$
$144 = (MP)^2$
$12 = MP$

25. $AB = BC$
$2x + 3 = x + 7$
$x = 4$
$EF = DE = 3x + 2$
$EF = 3(4) + 2 = 14$

29. If 2 consecutive angles of a quadrilateral are supplementary, the quadrilateral is a trapezoid.
Given: Quadrilateral $ABCD$
   with $\angle A$ is supp. to $\angle B$
Prove: $ABCD$ is a trapezoid

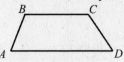

STATEMENTS	REASONS
1. Quad. $ABCD$ with $\angle A$ is supp. to $\angle B$	1. Given
2. $\overline{BC} \parallel \overline{AD}$	2. If 2 lines are cut by a trans. so that the interior $\angle$s on the same side of the trans. are supp., then these lines are $\parallel$.
3. $ABCD$ is a trapezoid	3. If a quad had 2 $\parallel$ sides, then the quad. is a trapezoid.

33. Given: $\overline{EF}$ is the median of trapezoid $ABCD$
Prove: $EF = \dfrac{1}{2}(AB + DC)$

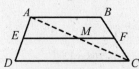

STATEMENTS	REASONS
1. $\overline{EF}$ is the median of trapezoid $ABCD$	1. Given
2. $\overline{AB} \parallel \overline{DC}$	2. Trapezoid has one pair of $\parallel$ sides.
3. $\overline{EF}$ is $\parallel$ to both $\overline{AB}$ and $\overline{DC}$	3. The median of a trap. is $\parallel$ to each base.
4. E is the midpoint of $\overline{AD}$ and F is the midpoint of $\overline{BC}$	4. The median of a trap. joins the midpoints of the nonparallel sides.
5. $\overline{AE} \cong \overline{ED}$	5. The midpoint of a segment forms 2 $\cong$ segments.
6. $\overline{AM} \cong \overline{MC}$	6. If 3 (or more) parallel lines intercept $\cong$ segments on one transversal, then they intercept $\cong$ segments on any transversal.
7. M is the midpoint of $\overline{AC}$	7. If a point divides a segment into 2 $\cong$ segments, then the point is the midpoint.
8. In $\triangle ADC$, $EM = \dfrac{1}{2}(DC)$ and in $\triangle ABC$, $MF = \dfrac{1}{2}(AB)$	8. The segment that joins the midpoints of two sides of a $\triangle$ is $\parallel$ to the third side and has a length equal to one-half the length of the third side.
9. $EM + MF = \dfrac{1}{2}(AB) + \dfrac{1}{2}(DC)$	9. Addition Property of Equality
10. $EM + MF = \dfrac{1}{2}(AB + DC)$	10. Distrubutive Property
11. $EF = EM + MF$	11. Segment Addition Postulate
12. $EF = \dfrac{1}{2}(AB + DC)$	12. Substitution

37. **a.** $AS = 3$ ft

b. $VD = 12$ ft

c. $CD = 13$ ft

d. $DE = \sqrt{73}$ ft

41. If $\angle A$ is supp to $\angle B$, then

$$\frac{x}{3} + 50 + \frac{x}{2} + 10 = 180$$

$$\frac{x}{3} + \frac{x}{2} = 120$$

Multiply by 6, the LCD

$2x + 3x = 720$

$5x = 720$

$x = 144$

If $\angle B$ is supp to $\angle C$, then

$$\frac{x}{3} + 50 + \frac{x}{5} + 50 = 180$$

$$\frac{x}{3} + \frac{x}{5} = 80$$

Multiply by 15, the LCD

$5x + 3x = 1200$

$8x = 1200$

$x = 150$

Possible values of x are 144 or 150.

CHAPTER REVIEW

1. A

2. S

3. N

4. S

5. S

6. A

7. A

8. A

9. A

10. N

11. S

12. N

13. $2(2x+3) + 2(5x-4) = 96$
$4x + 6 + 10x - 8 = 96$
$14x - 2 = 06$
$14x = 98$
$x = 7$
$AB = DC = 2(7) + 3 = 17$
$AD = BC = 5(7) - 4 = 31$

14. $2x + 6 + x + 24 = 180$
$3x + 30 = 180$
$3x = 150$
$x = 50$
$m\angle C = m\angle A = 2(50) + 6 = 106°$

15. The sides of a parallelogram measure 13 since $5^2 + 12^2 = (\text{side})^2$. Perimeter is 52.

16. $4x = 2x + 50$
$2x = 50$
$x = 25$
$m\angle M = 4(25) = 100°$
$m\angle P = 180 - 100 = 80°$

17. $\overline{PN}$

18. Kite

19. $m\angle G = m\angle F = 180 - 108 = 72°$
$m\angle E = 108°$

20. Median $= \frac{1}{2}(12.3 + 17.5)$
$= \frac{1}{2}(29.8)$
$= 14.9$

21. $15 = \frac{1}{2}(3x + 2 + 2x - 7)$
$30 = 5x - 5$
$35 = 5x$
$x = 7$
$MN = 3(7) + 2 = 23$
$PO = 2(7) - 7 = 7$

22. If $\overline{FJ} \cong \overline{FH}$ and M and N are their midpoints, then $FM = NH$ or
$2y + 3 = 5y - 9$
$-3y = -12$
$y = 4$
$FM = 2(4) + 3 = 11$
$FN = NH = 5(4) - 9 = 11$
$JH = 2(4) = 8$
The perimeter of $\triangle FMN = 26$.

23. Since M and N are midpoints, $\overline{MN} \parallel \overline{JH}$ and

$MN = \dfrac{1}{2} \cdot JH$. There fore, $MN = 6$,

$m\angle FMN = 80^\circ$ and $m\angle FNM = 40^\circ$.

24. Since M and N are midpoints, $MN = \dfrac{1}{2} \cdot JH$.

Therefore $x^2 + 6 = \dfrac{1}{2} \cdot 2x(x+2)$

$\qquad\qquad x^2 + 6 = x(x+2)$

$\qquad\qquad x^2 + 6 = x^2 + 2x$

$\qquad\qquad\qquad 6 = 2x$

$\qquad\qquad\qquad x = 3$

$MN = 15$

$JH = 30$

25. Given: $ABCD$ is a parallelogram

$\qquad\quad \overline{AF} \cong \overline{CE}$

Prove: $\overline{DF} \parallel \overline{EB}$

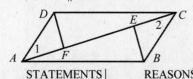

	STATEMENTS		REASONS
1.	$ABCD$ is a parallelogram	1.	Given
2.	$\overline{AD} \cong \overline{CB}$	2.	Opp. sides of a parallelogram are $\cong$.
3.	$\overline{AD} \parallel \overline{CB}$ parallelogram	3.	Opp. sides of a parallelogram are $\parallel$.
4.	$\angle 1 \cong \angle 2$	4.	If 2 $\parallel$ lines are cut by a trans., then the alt. int. $\angle$s are $\cong$.
5.	$\overline{AF} \cong \overline{CE}$	5.	Given
6.	$\triangle DAF \cong \triangle BCE$	6.	SAS
7.	$\angle DFA \cong \angle BEC$	7.	CPCTC
8.	$\overline{DF} \parallel \overline{EB}$	8.	If 2 lines are cut by a trans. so that alt. ex. $\angle$s are $\cong$, then the lines are $\parallel$.

26. Given: $ABEF$ is a rect., $BCDE$ is a rect.

$\qquad\quad \overline{FE} \cong \overline{ED}$

Prove: $\overline{AE} \cong \overline{BD}$ and $\overline{AE} \parallel \overline{BD}$

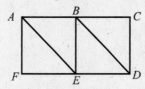

	STATEMENTS		REASONS
1.	$ABEF$ is a rect.	1.	Given
2.	$ABEF$ is a parallelogram	2.	A rect. is a parallelogram with a rt. $\angle$.
3.	$\overline{AF} \cong \overline{BE}$	3.	Opp. sides of a parallelogram are $\cong$.
4.	$BCDE$ is a rect.	4.	Given
5.	$\angle F$ and $\angle BED$ are rt. $\angle$s	5.	Same as (2)
6.	$\angle F \cong \angle BED$	6.	Any 2 rt. $\angle$s are $\cong$.
7.	$\overline{FE} \cong \overline{ED}$	7.	Given
8.	$\triangle AFE \cong \triangle BED$	8.	SAS
9.	$\overline{AE} \cong \overline{BD}$	9.	CPCTC
10.	$\angle AEF \cong \angle BDE$	10.	CPCTC
11.	$\overline{AE} \parallel \overline{BD}$	11.	If lines are cut by a trans. so that the corresp. $\angle$s are $\cong$, then the lines are $\parallel$.

27. Given: $\overline{DE}$ is a median in $\triangle ADC$

$\qquad\quad \overline{BE} \cong \overline{FD}$ and $\overline{EF} \cong \overline{FD}$

Prove: $ABCF$ is a parallelogram

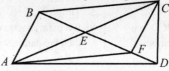

	STATEMENTS		REASONS
1.	$\overline{DE}$ is a median of $\triangle ADC$	1.	Given
2.	E is the midpoint of $\overline{AC}$	2.	Median of a $\triangle$ is a line segment drawn from a vertex to the midpoint of the opp. side.
3.	$\overline{AE} \cong \overline{EC}$	3.	Midpoint of a segment forms 2 $\cong$ segments.
4.	$\overline{BE} \cong \overline{FD}$ and $\overline{EF} \cong \overline{FD}$	4.	Given
5.	$\overline{BE} \cong \overline{EF}$	5.	Transitive Prop. for $\cong$
6.	$ABCF$ is a parallelogram	6.	If the diagonals of a quad. bisect each other then the quad is a parallelogram.

28. Given: △FAB ≅ △HCD
       △EAD ≅ △GCB
    Prove: ABCD is a parallelogram

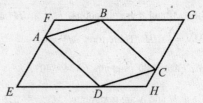

STATEMENTS	REASONS
1. △FAB ≅ △HCD	1. Given
2. $\overline{AB} \cong \overline{DC}$	2. CPCTC
3. △EAD ≅ △GCB	3. Given
4. $\overline{AD} \cong \overline{BC}$	4. CPCTC
5. ABCD is a parallelogram	5. If a quad. has both pairs of opp. sides ≅, then the quad is a parallelogram.

29. Given: ABCD is a parallelogram
       $\overline{DC} \cong \overline{BN}$
       ∠3 ≅ ∠4
    Prove: ABCD is a rhombus

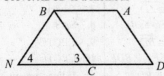

STATEMENTS	REASONS
1. ABCD is a parallelogram	1. Given
2. $\overline{DC} \cong \overline{BN}$	2. Given
3. ∠3 ≅ ∠4	3. Given
4. $\overline{BN} \cong \overline{BC}$	4. If 2 ∠s of a △ are ≅, then the sides opp. these ∠s are also ≅.
5. $\overline{DC} \cong \overline{BC}$	5. Transitive Prop. for ≅.
6. ABCD is a rhombus	6. If a parallelogram has 2 ≅ adj. sides, then the parallelogram is a rhombus

30. Given: △*TWX* is an isosceles with base $\overline{WX}$

$\overline{RY} \parallel \overline{WX}$

Prove: *RWXY* is an isosceles trapezoid

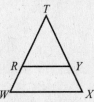

STATEMENTS	REASONS
1. △*TWX* is isosceles with base $\overline{WX}$	**1.** Given
2. ∠*W* ≅ ∠*X*	**2.** Base ∠s of an isos. △ are ≅.
3. $\overline{RY} \parallel \overline{WX}$	**3.** Given
4. ∠*TRY* ≅ ∠*W* and ∠*TYR* ≅ ∠*X*	**4.** If 2 ∥ lines are cut by a trans., then the corresp. ∠s are ≅.
5. ∠*TRY* ≅ ∠*TYR*	**5.** Transitive Prop. for ≅.
6. $\overline{TR} \cong \overline{TY}$	**6.** If 2 ∠s of a △ are ≅, then the side opp. these ∠s are also ≅.
7. $\overline{TW} \cong \overline{TX}$	**7.** Isosceles △ has 2 ≅ sides.
8. *TR* = *TY* and *TW* = *TX*	**8.** If 2 segments are ≅, then they are equal in length.
9. *TW* = *TR* + *RW* and *TX* = *TY* + *YX*	**9.** Segment-Addition Post.
10. *TR* + *RW* = *TY* + *YX*	**10.** Substitution
11. *RW* = *YX*	**11.** Subtraction Prop. of Eq.
12. $\overline{RW} \cong \overline{YX}$	**12.** If segments are ≅ in length, then they are ≅.
13. *RWXY* is an isosceles trapezoid.	**13.** If a quad. has one pair of ∥ sides and the non-parallel sides are ≅, then the quad. is an isos. trap.

31.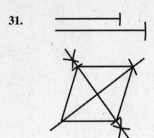

32. a. $\overline{AB} \perp \overline{BC}$

    b. $AC = 13$

33. a. $\overline{WY} \perp \overline{XZ}$

    b. $WY = 30$

34. a. Kites, rectangles, squares, rhombi, isosceles trapezoids

    b. Parallelograms, rectangles, squares, rhombi

35. a. Rhombus

    b. Kite

CHAPTER TEST

1. a. Congruent

    b. Supplementary

2. 18.8 cm

3. $CD = AB$. Let $x = AE$.
$$(AE)^2 + (DE)^2 = (AD)^2$$
$$x^2 + 4^2 = 5^2$$
$$x^2 = 9$$
$$x = 3$$
$$EB = AB - AE = 9 - 3 = 6$$

4. $m\angle S = 57°$
$$m\angle R = 180 - 57 = 123°$$
$\overline{VS}$ is longer.

5. $VT = 3x - 1$, $TS = 2x + 1$, $RS = 4(x - 2)$
$$3x - 1 = 4(x - 2)$$
$$x = 7$$

6. a. Kite

    b. Parallelogram

7. a. Altitude

    b. Rhombus

8. a. The line segments are parallel.

    b. $MN = \dfrac{1}{2}(BC)$

9. $MN = \dfrac{1}{2}(BC)$
$$7.6 = \dfrac{1}{2}(BC)$$
$$BC = 15.2 \text{ cm}$$

10. $MN = 3x - 11$
$$BC = 4x + 24$$
$$MN = \dfrac{1}{2}(BC)$$
$$3x - 11 = \dfrac{1}{2}(4x + 24)$$
$$x = 23$$

11. Let $x = AC$. $AD = 12$ and $DC = 5$
$$5^2 + 12^2 = x^2$$
$$x^2 = 169$$
$$x = 13$$
$$AC = 13$$
Or use the Pythagorean Triple (5, 12, 13).

12. a. $\overline{RV}$, $\overline{ST}$

    b. $\angle R$ and $\angle V$ (or $\angle S$ and $\angle T$)

13. $MN = \dfrac{1}{2}(RS + VT)$
$$= \dfrac{1}{2}(12.4 + 16.2)$$
$$= 14.3 \text{ in.}$$

14. $VT = 2x + 9$, $MN = 6x - 13$, $RS = 15$
$$MN = \dfrac{1}{2}(RS + VT)$$
$$6x - 13 = \dfrac{1}{2}(15 + 2x + 9)$$
$$6x - 13 = x + 12$$
$$5x = 25$$
$$x = 5$$

15. **S1.** Kite $ABCD$; $\overline{AB} \cong \overline{AD}$ and $\overline{BC} \cong \overline{DC}$

    R1. Given

    S3. $\overline{AC} \cong \overline{AC}$

    R4. SSS

    S5. $\angle B \cong \angle D$

    R5. CPCTC

16. **S1.** Trap. $ABCD$ with $\overline{AB} \parallel \overline{DC}$ and $\overline{AD} \cong \overline{BC}$

    R2. Congruent

    R3. Identity

    R4. SAS

    S5. $\overline{AC} \cong \overline{DB}$

17.　$x - 1 = y - 3$

Using $y = 2x - 4$, we get

$x - 1 = 2x - 4 - 3$

$x - 1 = 2x - 7$

$x = 6$;　$y = 2(6) - 4 = 8$;

$RS = 8$; $ST = 5$; $TV = 5$; $RV = 8$

Perimeter of $RSTV = 2(8) + 2(5) = 26$

Chapter 5: Similar Triangles

SECTION 5.1: Ratios, Rates and Proportions

1. **a.** $\dfrac{12}{15} = \dfrac{4}{5}$

   b. $\dfrac{12 \text{ inches}}{15 \text{ inches}} = \dfrac{4}{5}$

   c. $\dfrac{1 \text{ foot}}{18 \text{ inches}} = \dfrac{12 \text{ inches}}{18 \text{ inches}} = \dfrac{2}{3}$

   d. $\dfrac{1 \text{ foot}}{18 \text{ ounces}}$ is incommensurable

5. **a.** $12x = 36$
      $x = 3$

   b. $21x = 168$
      $x = 8$

9. **a.** $x^2 = 28$
      $x = \pm\sqrt{28} = \pm 2\sqrt{7} \approx \pm 5.29$

   b. $x^2 = 18$
      $x = \pm\sqrt{18} = \pm 3\sqrt{2} \approx \pm 4.24$

13. **a.** $\quad 3(x+1) = 2x^2$
      $\quad 3x + 3 = 2x^2$
      $2x^2 - 3x - 3 = 0$
      $a = 2; \ b = -3; \ c = -3$

   $x = \dfrac{-b \pm \sqrt{b^2 - 4ac}}{2a}$

   $x = \dfrac{-(-3) \pm \sqrt{(-3)^2 - 4(2)(-3)}}{2(2)}$

   $x = \dfrac{3 \pm \sqrt{9 + 24}}{4}$

   $x = \dfrac{3 \pm \sqrt{33}}{4} \approx 2.19 \text{ or } -0.69$

   b. $5(x+1) = 2x(x-1)$
      $5x + 1 = 2x^2 - 2x$
      $0 = 2x^2 - 7x - 5$
      $a = 2; \ b = -7; \ c = -5$

   $x = \dfrac{-b \pm \sqrt{b^2 - 4ac}}{2a}$

   $x = \dfrac{-(-7) \pm \sqrt{(-7)^2 - 4(2)(-5)}}{2(2)}$

   $x = \dfrac{7 \pm \sqrt{49 + 40}}{4}$

   $x = \dfrac{7 \pm \sqrt{89}}{4} \approx 4.19 \text{ or } -0.61$

17. $\dfrac{4 \text{ eggs}}{3 \text{ cups of milk}} = \dfrac{14 \text{ eggs}}{x \text{ cups of milk}}$
      $4x = 42$
      $x = \dfrac{42}{4} \text{ or } 10\dfrac{1}{2} \text{ cups of milk}$

21. **a.** $\quad \dfrac{BD}{AD} = \dfrac{AD}{DC}$
      $\quad \dfrac{6}{AD} = \dfrac{AD}{8}$
      $(AD)^2 = 48$
      $AD = \sqrt{48} = 4\sqrt{3} \approx 6.93$

   b. $\quad \dfrac{BD}{AD} = \dfrac{AD}{DC}$
      $\quad \dfrac{BD}{6} = \dfrac{6}{8}$
      $8(BD) = 36$
      $BD = \dfrac{36}{8} = 4\dfrac{1}{2}$

25. Let the first angle have measure x so that the complementary angle has measure $90 - x$. Then
      $\dfrac{x}{90 - x} = \dfrac{4}{5}$
      $5x = 4(90 - x)$
      $5x = 360 - 4x$
      $9x = 360$
      $x = 40; \ 90 - x = 50$
   The angles measure $40°$ and $50°$.
   Alternate method: Let the measures of the two angles be $4x$ and $5x$. Then
      $4x + 5x = 90$
      $9x = 90$
      $x = 10; \ 4x = 40; \ 5x = 50$
   The angles measure $40°$ and $50°$.

29. $\dfrac{7}{3} = \dfrac{6}{YZ}$
      $7 \cdot YZ = 18$
      $YZ = 2\dfrac{4}{7} \approx 2.57$

35. $\dfrac{1 \text{ in.}}{3 \text{ ft}} = \dfrac{x \text{ in.}}{12 \text{ ft}}$
      $3x = 12$
      $x = 4 \text{ in.}$
      $\dfrac{1 \text{ in.}}{3 \text{ ft}} = \dfrac{y \text{ in.}}{14 \text{ ft}}$
      $3y = 14$
      $y = 4\dfrac{2}{3} \text{ in.}$

   The blue print should be 4 in. by $4\dfrac{2}{3}$ in.

39. If $\dfrac{a}{b} = \dfrac{c}{d}$ where b • 0 and d • 0, add 1 to both

sides to get $\dfrac{a}{b} + 1 = \dfrac{c}{d} + 1$ or

$\dfrac{a}{b} + \dfrac{b}{b} = \dfrac{c}{d} + \dfrac{d}{d}$ or

$\dfrac{a+b}{b} = \dfrac{c+d}{d}$

SECTION 5.2: Similar Polygons

1. a. Congruent

   b. Proportional

5. a. $\triangle ABC \sim \triangle XTN$

   b. $\triangle ACB \sim \triangle NXT$

9. a. $m\angle N = 82°$

   b. $m\angle N = 42°$

   c. $\dfrac{NP}{RS} = \dfrac{NM}{RQ}$

   $\dfrac{NP}{7} = \dfrac{9}{6}$ or $\dfrac{3}{2}$

   $2 \cdot NP = 21$

   $NP = 10\dfrac{1}{2}$

   d. $\dfrac{MP}{QS} = \dfrac{MN}{RQ}$

   $\dfrac{12}{QS} = \dfrac{9}{6}$ or $\dfrac{3}{2}$

   $3 \cdot QS = 24$

   $QS = 8$

13. $\dfrac{HK}{KF} = \dfrac{HJ}{FG}$

   $\dfrac{4}{FG} = \dfrac{6}{8}$ or $\dfrac{3}{4}$

   $3 \cdot FG = 16$

   $FG = 5\dfrac{1}{3}$

17. $\dfrac{AB}{HJ} = \dfrac{BC}{JK}$

   $\dfrac{n}{n+3} = \dfrac{5}{10}$ or $\dfrac{1}{2}$

   $n+3 = 2n$

   $n = 3$

21. Let $BC = x$; then $CE = x$ and $CA = x+6$.

   $\dfrac{4}{x} = \dfrac{6}{x+6}$

   $6x = 4(x+6)$

   $6x = 4x + 24$

   $2x = 24$

   $x = 12;\ BC = 12$

25. Quadrilateral $MNPQ \sim$ quadrilateral $WXYZ$

   $\dfrac{6}{9} = \dfrac{50}{n}$

   $6n = 9 \cdot 50$

   $6n = 450$

   $n = 75$

29. Let $x =$ the boy's height.

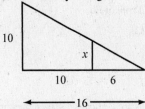

   $\dfrac{6}{x} = \dfrac{16}{10}$ or $\dfrac{8}{5}$

   $8x = 30$

   $x = \dfrac{30}{8}$ or $3\dfrac{3}{4}$, the boy is 3 ft 9 in.

33. No. The sides of quadrilateral $ABCD$ are not proportional to quadrilateral $DCFE$.

37. In rhombus ARST, if AR = RS = TS = AT = x, then TC = 6 • x and RB = 10 • x. The

   proportion to use is $\dfrac{TC}{TS} = \dfrac{RS}{BR}$ or

   $\dfrac{6-x}{x} = \dfrac{x}{10-x}$

   $x^2 = (6 \bullet x)(10 \bullet x)$

   $x^2 = 60 \bullet 16x + x^2$

   $16x = 60$

   $x = 3.75$

SECTION 5.3: Proving Triangles Similar

1. CASTC

5. *SSS ~*

9. *SAS ~*

13. **1.** Given

    2. Definition of midpoint

    3. If a line segment joins the midpoints of two sides of a $\triangle$, its length is $\frac{1}{2}$ the length of the third side.

    4. Division Prop. of Eq.

    5. Substitution

    6. *SSS ~*

17. **1.** $\angle H \cong \angle F$

    2. If two $\angle$ s are vertical $\angle$ s, then they are $\cong$.

    S3. $\triangle HJK \sim \triangle FGK$

    R3. AA

21. **S1.** $\overline{RS} \parallel \overline{UV}$

    R1. Given

    2. 2 $\parallel$ lines are cut by a transversal, alternate interior $\angle$ s are $\cong$.

    3. $\triangle RST \sim \triangle VUT$

    S4. $\dfrac{RT}{VT} = \dfrac{RS}{VU}$

    R4. CSSTP

25. Let $DB = x$; then $AB = DB + AD = x + 4$.

$$\frac{AC}{DE} = \frac{AB}{DB}$$
$$\frac{10}{8} = \frac{x+4}{x}$$
$$8(x+4) = 10x$$
$$8x + 32 = 10x$$
$$2x = 32$$
$$x = 16$$
$$DB = 16$$

29. Since $\triangle ABF \sim \triangle CBD$, then.

$$m\angle C + m\angle B + m\angle AFB = 180°$$
$$45° + x + 4x = 180°$$
$$5x = 135°$$
$$x = 27°$$

33. "The lengths of the corresponding altitudes of similar triangles have the same ratio as the lengths of any pair of corresponding sides."

Given: $\triangle DEF \sim \triangle MNP$

    $\overline{DG}$ and $\overline{MQ}$ are altitudes

Prove: $\dfrac{DG}{MQ} = \dfrac{DE}{MN}$

STATEMENTS	REASONS
1. $\triangle DEF \sim \triangle MNP$ $\overline{DG}$ and $\overline{MQ}$ are altitudes.	**1.** Given
2. $\overline{DG} \perp \overline{EF}$ and $\overline{MQ} \perp \overline{NP}$	**2.** An altitude is a segment drawn from a vertex $\perp$ to the opposite side.
3. $\angle DGE$ and $\angle MQN$ are rt. $\angle$s	**3.** $\perp$ lines form a rt. $\angle$.
4. $\angle DGE \cong \angle MQN$	**4.** Right $\angle$s are $\cong$.
5. $\angle E \cong \angle N$	**5.** If two $\triangle$s are $\sim$ then the corresponding $\angle$s are $\cong$.
6. $\triangle DGE \sim \triangle MQN$	**6.** AA
7. $\dfrac{DG}{MQ} = \dfrac{DE}{MN}$	**7.** Corresp. sides of $\sim$ $\triangle$s are proportional.

37. $\dfrac{XY}{YZ} = \dfrac{TY}{YW} = \dfrac{XT}{WZ}$

$$\frac{120}{40} = \frac{XT}{50}$$
$$3 = \frac{XT}{50}$$
$$XT = 150 \text{ feet}$$

SECTION 5.4: The Pythagorean Theorem

1. $\triangle RST \sim \triangle RVS \sim \triangle SVT$

5. $\dfrac{RV}{6} = \dfrac{6}{8}$
$8 \cdot RV = 36$
$RV = 4.5$

9. a. $(DF)^2 = (DE)^2 + (EF)^2$
$17^2 = 15^2 + (EF)^2$
$289 = 225 + (EF)^2$
$(EF)^2 = 64$
$EF = 8$
Or, $(8, 15, 17)$ is a Pythagorean Triple; therefore $EF = 8$.

b. $(DF)^2 = (DE)^2 + (EF)^2$
$12^2 = \left(8\sqrt{2}\right)^2 + (EF)^2$
$144 = 128 + (EF)^2$
$(EF)^2 = 16$
$EF = 4$

13. Let c be the longest side.

a. $5^2 = 3^2 + 4^2$
Right $\triangle$

b. $6^2 < 4^2 + 5^2$
Acute $\triangle$

c. $\left(\sqrt{7}\right)^2 = 2^2 + \left(\sqrt{3}\right)^2$
Right $\triangle$

d. No $\triangle$

17. Let x = the length of rope needed.

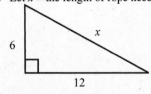

$x^2 = 6^2 + 12^2$
$x^2 = 36 + 144$
$x^2 = 180$
$x = \sqrt{180} = 6\sqrt{5} \approx 13.4$ meters

21. Let x = the length of the rectangle.

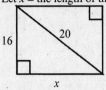

$20^2 = 16^2 + x^2$
$400 = 256 + x^2$
$x^2 = 144$
$x = 12$ cm
Or $(12, 16, 20)$ is a multiple of the Pythagorean Triple $(3, 4, 5)$.

25. Since the diagonals of a rhombus are perpendicular and since they bisect each other, one side of the right triangle has length 9 and the hypotenuse has length 12.
Let x represent the length of the other side of the right triangle.
$12^2 = x^2 + 9^2$
$144 = x^2 + 81$
$x^2 = 63$
$x = \sqrt{63} = 3\sqrt{7}$
The length of the other diagonal is $6\sqrt{7} \approx 15.87$ in.

29. In rt. $\triangle ACB$, $AC = 8$ using the triple $(8, 15, 17)$. Since M and N are midpoints, $\overline{MN} \parallel \overline{AC}$. Since $\overline{MN} \parallel \overline{AC}$, $MN = \dfrac{1}{2} \cdot AC$. Therefore, $MN = 4$.

33. In rt. $\triangle RST$, if $RS = 6$ and $ST = 8$, then $RT = 10$ using the Pythagorean Triple $(6, 8, 10)$. In rt. $\triangle RTU$, $RT = 10$ and $RU = 15$; let $UT = x$ and use the Pythagorean Theorem.
$15^2 = 10^2 + x^2$
$225 = 100 + x^2$
$x^2 = 125$
$x = \sqrt{125} = 5\sqrt{5} \approx 11.18$

37.

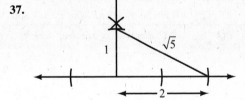

41. $TS = 13$ using the Pythagorean Triple (5, 12, 13).

$$(RT)^2 = (RS)^2 + (TS)^2$$
$$(RT)^2 = 13^2 + 13^2$$
$$(RT)^2 = 169 + 169$$
$$(RT)^2 = 338$$
$$RT = \sqrt{338} = 13\sqrt{2} \approx 18.38$$

SECTION 5.5: Special Right Triangles

1. a. $BC = a$

   b. $AB = \sqrt{a^2 + a^2} = a\sqrt{2}$

5. $YZ = 8$ and $XY = 8\sqrt{2} \approx 11.31$

9. $DF = 5\sqrt{3} \approx 8.66$ and $FE = 10$

13. In right $\triangle HLK$, if $m\angle HKL = 30°$ and
   $LK = 6\sqrt{3}$, then $HL = 6$ and $HK = 12$. $MK = 6$
   since the diagonals of a rectangle bisect each
   other.

17. $RS = 6$ and $RT = 6\sqrt{3} \approx 10.39$.

21. From vertex Z, draw an altitude to $\overline{XY}$; call the
   altitude $\overline{ZW}$. In the 30-60-90 $\triangle$, $WX = 6$ and
   $ZW = 6\sqrt{3}$. In the 45-45-90 $\triangle$, $\overline{YW} = 6\sqrt{3}$.
   $XY = YW + WX = 6\sqrt{3} + 6 \approx 16.39$.

25. 60°; $200^2 + x^2 = 400^2$
$$x^2 = 400^2 - 200^2$$
$$x^2 = 120,000$$
$$x \approx 346$$
   The jogger travels $(200 + 346) - 400 = 146$ feet
   further.

29. Since $\triangle MNQ$ is equiangular and $\overrightarrow{NR}$ bisects
   $\angle MNQ$ and $\overrightarrow{QR}$ bisects $\angle MQN$,
   $m\angle RQN = 30° = m\angle RNQ$. From R, draw an
   altitude to $\overline{NQ}$. Name the altitude $\overline{RP}$.
   $NR = RQ = 6$. In 30-60-90 $\triangle RPQ$, $RP = 3$ and
   $PQ = 3\sqrt{3}$. NQ therefore equals $6\sqrt{3} \approx 10.39$.

33. Draw in altitude $\overline{CD}$. In right $\triangle CDB$, if
   $BC = 12$, then $CD = 6$ and $DB = 6\sqrt{3}$. In right
   $\triangle ACD$, if $CD = 6$, then $AD = 6$ and
   $AC = 6\sqrt{2}$. $AB = 6 + 6\sqrt{3} \approx 16.39$.

37. Draw $\overline{XV}$. $YV = VZ = 2$; $XZ = 5$.

Using the Pythagorean Theorem on $\triangle XYV$,

$$XV = \sqrt{3^2 + 2^2} = \sqrt{13}$$

Let $VW = x$; $WZ = y$ and $XW = 5 - y$

In $\triangle XVW$, $x^2 + (5 - y)^2 = (\sqrt{13})^2$

In $\triangle VWZ$, $x^2 + y^2 = 2^2$ or $x^2 = 4 - y^2$.

By substitution and squaring $(5 - y)$,

$$4 - y^2 + 25 - 10y + y^2 = 13$$
$$29 - 10y = 13$$
$$-10y = -16$$
$$y = 1.6$$

Using $x^2 = 4 - y^2$

$$x^2 = 4 - (1.6)^2$$
$$x^2 = 4 - 2.56 = 1.44$$
$$x = 1.2; \; VW = 1.2$$

SECTION 5.6: Segments Divided Proportionately

1. Let $5x =$ the amount of ingredient A;
$4x =$ amount of ingredient B;
$6x =$ amount of ingredient C.
$5x + 4x + 6x = 90$
$15x = 90$
$x = 6$
30 ounces of ingredient A;
24 ounces of ingredient B;
36 ounces of ingredient C.

5. Let $EF = x$, $FG = y$, and $GH = z$
$AD = 5 + 4 + 3 = 12$ so that

$\dfrac{AB}{AD} = \dfrac{EF}{EH}$ $\qquad$ $\dfrac{BC}{AD} = \dfrac{FG}{EH}$ $\qquad$ $\dfrac{CD}{AD} = \dfrac{GH}{EH}$

$\dfrac{5}{12} = \dfrac{x}{10}$ $\qquad$ $\dfrac{4}{12} = \dfrac{y}{10}$ $\qquad$ $\dfrac{3}{12} = \dfrac{z}{10}$

$12x = 50$ $\qquad$ $12y = 40$ $\qquad$ $12z = 30$

$x = \dfrac{50}{12} = 4\dfrac{1}{6}$ $\quad$ $y = \dfrac{40}{12} = 3\dfrac{1}{3}$ $\quad$ $z = \dfrac{30}{12} = 2\dfrac{1}{2}$

$EF = 4\dfrac{1}{6}$ $\qquad$ $FG = 3\dfrac{1}{3}$ $\qquad$ $GH = 2\dfrac{1}{2}$

9. Let $EC = x$.
$\dfrac{5}{12} = \dfrac{7}{x}$
$5x = 84$
$x = \dfrac{84}{5}$ or $16\dfrac{4}{5}$; $EC = 16\dfrac{4}{5}$

13. a. No

b. Yes

17. Let $NP = x = MQ$.
$\dfrac{x}{12} = \dfrac{8}{x}$
$x^2 = 96$
$x = \sqrt{96} = 4\sqrt{6}$
$NP = 4\sqrt{6} \approx 9.80$

23. If $RS = 6$ and $RT = 12$, then $\triangle RST$ is a
30-60-90 $\triangle$ and $ST = 6\sqrt{3}$. Let $SV = x$. Then
$VT = 6\sqrt{3} - x$.
$\dfrac{6}{12} = \dfrac{x}{6\sqrt{3} - x}$
$36\sqrt{3} - 6x = 12x$
$36\sqrt{3} = 18x$
$x = 2\sqrt{3}$
$SV = 2\sqrt{3} \approx 3.46$
$VT = 6\sqrt{3} - 2\sqrt{3} = 4\sqrt{3} \approx 6.93$

27. a. True

b. True

31. 1. Given

2. Means-Extremes Property

3. Addition Property of Equality

4. Distributive Property

5. Means-Extremes Property

6. Substitution

35. Given: $\triangle XYZ$; $\overrightarrow{YW}$ bisects $\angle XYZ$
$\overline{WX} \cong \overline{WZ}$
Prove: $\triangle XYZ$ is isosceles

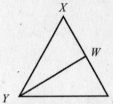

STATEMENTS	REASONS
1. $\triangle XYZ$; $\overrightarrow{YW}$ bisects $\angle XYZ$	1. Given
2. $\dfrac{YX}{YZ} = \dfrac{WX}{WZ}$	2. If a ray bisects one $\angle$ of a $\triangle$, then it divides the opposite side into segments that are proportional to the two sides which form that angle.
3. $\overline{WX} \cong \overline{WZ}$	3. Given
4. $WX = WZ$	4. If two segments are $\cong$, then their measures are equal. into four $\cong$ $\angle$s.
5. $\dfrac{YX}{YZ} = \dfrac{WX}{WX} = 1$	5. Substitution
6. $YX = YZ$	6. Means-Extremes Property
7. $\overline{YX} \cong \overline{YZ}$	7. If 2 segments are equal in measure, they are $\cong$.
8. $\triangle XYZ$ is isosceles	8. If 2 sides of a $\triangle$ are $\cong$, then the $\triangle$ is isosceles.

39. **a.** Let $CD = x$. Then $DB = 5 - x$.

$$\frac{CD}{CA} = \frac{DB}{BA}$$

$$\frac{x}{4} = \frac{5-x}{6}$$

$$4(5-x) = 6x$$
$$20 - 4x = 6x$$
$$10x = 20$$
$$x = 2$$
$$CD = 2$$
$$DB = 5 - 2 = 3$$

b. Let $CE = x$. Then $EA = 4 - x$.

$$\frac{CE}{BC} = \frac{EA}{BA}$$

$$\frac{x}{5} = \frac{4-x}{6}$$

$$5(4-x) = 6x$$
$$20 - 5x = 6x$$
$$11x = 20$$
$$x = \frac{20}{11}$$
$$CE = \frac{20}{11}$$
$$EA = 4 - \frac{20}{11} = \frac{24}{11}$$

c. Let $BF = x$. Then $FA = 6 - x$.

$$\frac{BF}{BC} = \frac{FA}{CA}$$

$$\frac{x}{5} = \frac{6-x}{4}$$

$$5(6-x) = 4x$$
$$30 - 5x = 4x$$
$$9x = 30$$
$$x = \frac{10}{3}$$
$$BF = \frac{10}{3}$$
$$FA = 6 - \frac{10}{3} = \frac{8}{3}$$

d. $\dfrac{BD}{DC} \cdot \dfrac{CE}{EA} \cdot \dfrac{AF}{FB} = \dfrac{3}{2} \cdot \dfrac{\frac{20}{11}}{\frac{24}{11}} \cdot \dfrac{\frac{8}{3}}{\frac{10}{3}} = \dfrac{3}{2} \cdot \dfrac{20}{24} \cdot \dfrac{8}{10} = 1$

CHAPTER REVIEW

1. False

2. True

3. False

4. True

5. True

6. False

7. True

8. **a.** $x^2 = 18$
$$x = \pm\sqrt{18} = \pm 3\sqrt{2} \approx \pm 4.24$$

b. $7(x-5) = 3(2x-3)$
$$7x - 35 = 6x - 9$$
$$x = 26$$

c. $6(x+2) = 2(x+4)$
$$6x + 12 = 2x + 8$$
$$4x = -4$$
$$x = -1$$

d. $7(x+3) = 5(x+5)$
$$7x + 21 = 5x + 25$$
$$2x = 4$$
$$x = 2$$

e. $(x-2)(x-1) = (x-5)(2x+1)$
$$x^2 - 3x + 2 = 2x^2 - 9x - 5$$
$$x^2 - 6x - 7 = 0$$
$$(x-7)(x+1) = 0$$
$$x = 7 \text{ or } x = -1$$

f. $\quad 5x(x+5) = 9(4x+4)$
$$5x^2 + 25x = 36x + 36$$
$$5x^2 - 11x - 36 = 0$$
$$(5x+9)(x-4) = 0$$
$$x = -\frac{9}{5} \text{ or } x = 4$$

g. $(x-1)(3x-2) = 10(x+2)$
$$3x^2 - 5x + 2 = 10x + 20$$
$$3x^2 - 15x - 18 = 0$$
$$x^2 - 5x - 6 = 0$$
$$(x-6)(x+1) = 0$$
$$x = 6 \text{ or } x = -1$$

h. $(x+7)(x-2) = 2(x+2)$
$$x^2 + 5x - 14 = 2x + 4$$
$$x^2 + 3x - 18 = 0$$
$$(x+6)(x-3) = 0$$
$$x = -6 \text{ or } x = 3$$

9. Let $x = $ cost of the six containers.

$$\frac{4}{2.52} = \frac{6}{x}$$
$$4x = 15.12$$
$$x = 3.78$$

The six containers cost $3.78.

10. Let $x = $ the number of packages you can buy for $2.25.

$$\frac{2}{0.69} = \frac{x}{2.25}$$
$$0.69x = 4.50$$
$$x = \frac{450}{69} = 6\frac{12}{23}$$

With $2.25, you can buy 6 packages of M&M's.

11. Let x = cost of the rug that is 12 square meters.

$$\frac{20}{132} = \frac{12}{x}$$

$$20x = 1584$$

$$x = 79.20$$

The 12 square meters rug will cost $79.20.

12. Let the measure of the sides of the quadrilateral be $2x$, $3x$, $5x$ and $7x$.

$$2x + 3x + 5x + 7x = 68$$

$$17x = 68$$

$$x = 4$$

The length of the sides are 8, 12, 20 and 28.

13. Let the width of the similar rectangle be x.

$$\frac{18}{12} = \frac{27}{x}$$

$$\frac{3}{2} = \frac{27}{x}$$

$$3x = 54$$

$$x = 18$$

The width of the similar rectangle is 18.

14. Let x and y be the lengths of the other two sides.

$$\frac{6}{15} = \frac{8}{x} = \frac{9}{y}$$

$$\frac{2}{5} = \frac{8}{x} \quad \text{and} \quad \frac{2}{5} = \frac{9}{y}$$

$$2x = 40 \qquad 2y = 45$$

$$x = 20 \qquad y = \frac{45}{2} = 22\frac{1}{2}$$

The other two sides have lengths 20 and $22\frac{1}{2}$.

15. Let the measure of the angle be x; the measure of the supplement would be $180 - x$; the measure of the complement would be $90 - x$.

$$\frac{180 - x}{90 - x} = \frac{5}{2}$$

$$2(180 - x) = 5(90 - x)$$

$$360 - 2x = 450 - 5x$$

$$3x = 90$$

$$x = 30$$

The measure of the supplement is 150°.

16. **a.** SSS ~

   b. AA

   c. SAS ~

   d. SSS ~

17. Given: $ABCD$ is a parallelogram;
    $\overline{DB}$ intersects $\overline{AE}$ at pt. F

   Prove: $\dfrac{AF}{EF} = \dfrac{AB}{DE}$

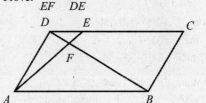

STATEMENTS	REASONS
1. $ABCD$ is a parallelogram $\overline{DB}$ intersects $\overline{AE}$ at pt. F	1. Given
2. $\overline{DC} \parallel \overline{AB}$	2. Opp. sides of a parallelogram are $\parallel$.
3. $\angle CDB \cong \angle ABD$	3. If 2 $\parallel$ lines are cut by a trans., then the alt. int. $\angle$s $\cong$.
4. $\angle DEF \cong \angle BAF$	4. Same as (3).
5. $\triangle DFE \sim \triangle BFA$	5. AA
6. $\dfrac{AF}{EF} = \dfrac{AB}{DE}$	6. Corresp. sides of $\sim$ $\triangle$s are proportional.

18. Given: $\angle 1 \cong \angle 2$

   Prove: $\dfrac{AB}{AC} = \dfrac{BE}{CD}$

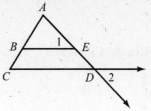

STATEMENTS	REASONS
1. $\angle 1 \cong \angle 2$	1. Given
2. $\angle ADC = \angle 2$	2. If 2 lines intersect, then the vertical formed are $\cong$.
3. $\angle ADC \cong \angle 1$	3. Transitive Prop. for Congruence.
4. $\angle A \cong \angle A$	4. Identity.
5. $\triangle BAE \sim \triangle CAD$	5. AA
6. $\dfrac{AB}{AC} = \dfrac{BE}{CD}$	6. Corresp. sides of $\sim$ $\triangle$s are proportional.

19. Since the △ s are ~ , $m\angle A = m\angle D$.

$50 = 2x + 40$

$10 = 2x$

$x = 5$

$m\angle D = 2(5) + 40 = 50°$

$m\angle E = 33°$

$m\angle F = 180 - 33 - 50 = 97°$

20. With $\angle B \cong \angle F$, and $\angle C \cong \angle E$,

$\triangle ABC \sim \triangle DFE$. It follows that

$\dfrac{AB}{DF} = \dfrac{AC}{DE} = \dfrac{BC}{FE}$

Substituting in gives

$\dfrac{AB}{2} = \dfrac{9}{3}$ or $\dfrac{3}{1}$ and $\dfrac{9}{3}$ or $\dfrac{3}{1} = \dfrac{BC}{4}$

$AB = 6$ and $BC = 12$.

21. $\dfrac{BD}{AD} = \dfrac{BE}{EC}$; let $AB = x$

$\dfrac{6}{x} = \dfrac{8}{4}$ or $\dfrac{2}{1}$

$2x = 6$

$x = 3$

22. $\dfrac{BD}{BA} = \dfrac{DE}{AC}$; let $AC = x$; $BA = 12$

$\dfrac{8}{12} = \dfrac{2}{3} = \dfrac{3}{x}$

$2x = 9$

$x = 4\dfrac{1}{2}$

$AC = 4\dfrac{1}{2}$

23. $\dfrac{BD}{BA} = \dfrac{BE}{BC}$; let $BC = x$; $BD = 8$

$\dfrac{8}{10}$ or $\dfrac{4}{5} = \dfrac{5}{x}$

$4x = 25$

$x = 6\dfrac{1}{4}$

$BC = 6\dfrac{1}{4}$

24. Since $\overrightarrow{GJ}$ bisects $\angle FGH$, we can write the

proportion $\dfrac{FJ}{FG} = \dfrac{JH}{GH}$; let $JH = x$.

$\dfrac{7}{10} = \dfrac{x}{8}$

$10x = 56$

$x = \dfrac{56}{10} = 5\dfrac{3}{5}$

$JH = 5\dfrac{3}{5}$

25. Since $\overrightarrow{GJ}$ bisects $\angle FGH$, and $GF : GH = 1 : 2$,

we can write the proportion $\dfrac{GF}{GH} = \dfrac{FJ}{JH}$; let

$JH = x$.

$\dfrac{1}{2} = \dfrac{5}{x}$

$x = 10$

$JH = 10$

26. Since $\overrightarrow{GJ}$ bisects $\angle FGH$, we can write the

proportion $\dfrac{FG}{GH} = \dfrac{FJ}{JH}$; let $FJ = x$ and

$JH = 15 - x$.

$\dfrac{8}{12} = \dfrac{x}{15 - x}$

$\dfrac{2}{3} = \dfrac{x}{15 - x}$

$2(15 - x) = 3x$

$30 - 2x = 3x$

$30 = 5x$

$x = 6$

$FJ = 6$

27. Let $MK = x$; then $\dfrac{MK}{HJ} = \dfrac{EM}{FH}$.

$\dfrac{x}{5} = \dfrac{6}{10}$ or $\dfrac{3}{5}$

$x = 3$

$MK = 3$

Let $EO = y$ and $OM = 6 - y$.

Then $\dfrac{y}{2} = \dfrac{6 - y}{8}$

$8y = 2(6 - y)$

$8y = 12 - 2y$

$10y = 12$

$y = \dfrac{12}{10} = 1\dfrac{1}{5}$

$EO = 1\dfrac{1}{5}$

$EK = 9$

28. If a line bisects one side of a triangle and is parallel to a second side, then it bisects the third side.

Given: $\overleftrightarrow{DE}$ bisects $\overline{AC}$

$\overleftrightarrow{DE} \parallel \overline{BC}$

Prove: $\overleftrightarrow{DE}$ bisects $\overline{AB}$

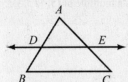

STATEMENTS	REASONS
1. $\overleftrightarrow{DE}$ bisects $\overline{AC}$	1. Given
2. $AE = EC$	2. Bisecting a segment forms 2 segments of equal measure.
3. $\overleftrightarrow{DE} \parallel \overline{BC}$	3. Given
4. $\dfrac{AD}{DB} = \dfrac{AE}{EC}$	4. If a line is parallel to one side of a triangle and intersects the other sides, then it divides these sides proportionally.
5. $AD \cdot EC = DB \cdot AE$	5. Means-Extremes Prop.
6. $AD = DB$	6. Division Prop. of Eq.
7. $\overleftrightarrow{DE}$ bisects $\overline{AB}$	7. If a segment has been divided into 2 segments of equal measure, the segment has been bisected.

29. The diagonals of a trapezoid divide themselves proportionally.

Given: $ABCD$ is a trapezoid with $\overline{BC} \parallel \overline{AD}$ and diagonals $\overline{BD}$ and $\overline{AC}$

Prove: $\dfrac{BE}{ED} = \dfrac{EC}{AE}$

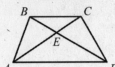

STATEMENTS	REASONS
1. $ABCD$ is a trapezoid with $\overline{BC} \parallel \overline{AD}$ and diagonals $\overline{BD}$ and $\overline{AC}$.	1. Given
2. $\angle CBE \cong \angle ADE$ and $\angle BCE \cong \angle DAE$	2. If 2 $\parallel$ lines are cut by a trans., then the alt. int. $\angle$s are $\cong$.
3. $\triangle BCE \cong \triangle DAE$	3. AA
4. $\dfrac{BE}{ED} = \dfrac{EC}{AE}$	4. Corresponding sides of $\sim$ $\triangle$s are proportional.

30. a. $\dfrac{BD}{AD} = \dfrac{AD}{DC}$

Let $DC = x$.

$\dfrac{3}{5} = \dfrac{5}{x}$

$3x = 25$

$x = \dfrac{25}{3}$

$DC = 8\dfrac{1}{3}$

b. $\dfrac{DC}{AC} = \dfrac{AC}{BC}$

Let $BD = x$ and $BC = x + 4$

$\dfrac{4}{5}$ or $\dfrac{2}{3} = \dfrac{10}{x+4}$

$2(x+4) = 50$

$2x + 8 = 50$

$2x = 42$

$x = 21$

$BD = 21$

c. $\dfrac{BD}{BA} = \dfrac{BA}{BC}$

Let $BA = x$.

$\dfrac{2}{x} = \dfrac{x}{6}$

$x^2 = 12$

$x = \sqrt{12} = 2\sqrt{3}$

$BA = 2\sqrt{3} \approx 3.46$

d. $\dfrac{DA}{AC} = \dfrac{AC}{BC}$

Let $DC = x$ and $BC = x + 3$.

$\dfrac{x}{3\sqrt{2}} = \dfrac{3\sqrt{2}}{x+3}$

$x(x+3) = 18$

$x^2 + 3x - 18 = 0$

$(x+6)(x-3) = 0$

$x = -6$ or $x = 3$; reject $x = -6$

$DC = 3$.

31. a. $\dfrac{AD}{BD} = \dfrac{BD}{DC}$

Let $DC = x$.

$\dfrac{9}{12}$ or $\dfrac{3}{4} = \dfrac{12}{x}$

$3x = 48$

$x = 16$

$DC = 16$

b. $\dfrac{DC}{BC} = \dfrac{BC}{AC}$

Let $AD = x$ and $AC = x + 5$.

$\dfrac{5}{15}$ or $\dfrac{1}{3} = \dfrac{15}{x+5}$

$x + 5 = 45$

$x = 40$

$AD = 40$.

c. $\dfrac{AD}{AB} = \dfrac{AB}{AC}$

Let $AB = x$ and $AC = 10$.

$\dfrac{2}{x} = \dfrac{x}{10}$

$x^2 = 20$

$x = \sqrt{20} = 2\sqrt{5}$

$AB = 2\sqrt{5} \approx 4.47$

d. $\dfrac{AD}{AB} = \dfrac{AB}{AC}$

Let $AD = x$ and $AC = x + 2$.

$\dfrac{x}{2\sqrt{6}} = \dfrac{2\sqrt{6}}{x+2}$

$x(x+2) = 24$

$x^2 + 2x - 24 = 0$

$(x+6)(x-4) = 0$

$x = -6$ or $x = 4$; reject $x = -6$

$AD = 4$.

32. a. $x = 30$. Since the leg is half of the hypotenuse, the angle opposite the leg must be 30°.

b. Half of the base is 10. In the right $\triangle$, 1 side has length 10 and the hypotenuse has length 26. The other side has length 24 since (10, 24, 26) is a multiple of (5, 12, 13).

c. $x^2 = 12^2 + 16^2$

$x^2 = 144 + 256$

$x^2 = 400$

$x = 20$

or (12, 16, 20) is a multiple of (3, 4, 5).

d. The unknown length of the right $\triangle$ is 8 using the Triple (8, 15, 17).

$x = 16$.

33. In rect. *ABCD*, $BC = 24$ and since *E* is a midpoint, $BE = 12$ and $EC = 12$. $CD = 16$ and $FD = 7$. There are three right triangles for which the Pythagorean Triples apply. $AE = 20$ using (12, 16, 20) which is a multiple of (3, 4, 5). $EF = 15$ using (9, 12, 15) which is also a multiple of (3, 4, 5). $AF = 25$ using the Triple (7, 24, 25).

34. In a square there are two 45-45-90 $\triangle$ s. If the length of the side of the square is 4 inches, then the length of the diagonal is $4\sqrt{2} \approx 5.66$ inches.

35. In a square there are two 45°-45°-90° $\triangle$ s. If the length of the diagonal is 6, then $6 = a\sqrt{2}$. Solving for *a* gives

$a = \dfrac{6}{\sqrt{2}} = \dfrac{6\sqrt{2}}{2} = 3\sqrt{2}$

Hence, the length of a side is $3\sqrt{2} \approx 4.24$ cm.

36. Since the diagonals of a rhombus are perpendicular and bisect each other, there are 4 right triangles formed whose sides are the lengths 24 cm and 7 cm. The hypotenuse must then have a length of 25. Since they hypotenuse of a right triangle is the side of the rhombus, the side has length 25 cm.

37. The altitude to one side of an equilateral triangle divides it into two 30°-60°-90° $\triangle$ s.

The altitude is the side opposite the 60 degree angle and is equal in length to one-half the length of the hypotenuse times $\sqrt{3}$. Hence, the altitude has length $5\sqrt{3} \approx 8.66$ in.

38. The altitude to one side of an equilateral triangle divides it into two 30°-60°-90° △ .

The altitude is the side opposite the 60 degree angle and is equal in length to one-half the length of the hypotenuse times $\sqrt{3}$. If H represents the length of the hypotenuse, then

$$6 = \frac{1}{2} \cdot H \cdot \sqrt{3}$$

$$12 = H \cdot \sqrt{3}$$

$$H = \frac{12}{\sqrt{3}} \cdot \frac{\sqrt{3}}{\sqrt{3}} = \frac{12\sqrt{3}}{3} = 4\sqrt{3}$$

The length of the sides of the △ is $4\sqrt{3} \approx 6.93$ in.

39. The altitude to the side of length 14 separates it into two parts the lengths of these are given by x and $14 - x$.

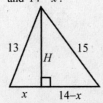

If the length of the altitude is H, we can use the Pythagorean Theorem on the two right triangles to get $x^2 + H^2 = 13^2$ and $(14 - x)^2 + H^2 = 15^2$. Subtracting the first equation from the second, we have

$$196 - 29x + x^2 + H^2 = 225$$

$$\underline{ x^2 + H^2 = 169}$$

$$196 - 28x = 56$$

$$-28x = -140$$

$$x = 5$$

Now we use x to find H.

$x^2 + H^2 = 13^2$ becomes

$$5^2 + H^2 = 169$$

$$H^2 = 144$$

$$H = 12$$

The length of the altitude is 12 cm.

40. a. Let the length of the hypotenuse common to both △ s be H.

$$\frac{1}{2} \cdot H \cdot \sqrt{3} = 9\sqrt{3}$$

$$H \cdot \sqrt{3} = 18\sqrt{3}$$

$$H = 18$$

$$y = \frac{1}{2} \cdot 18 = 9$$

$$x = 9\sqrt{2} \approx 12.73$$

b. $y = 6$ using the Triple (6, 8, 10). Since the length of the altitude to the hypotenuse is 6, 6 is the geometric mean for x and 8. That is,

$$\frac{x}{6} = \frac{6}{8}$$

$$8x = 36$$

$$x = \frac{36}{8} = \frac{9}{2} = 4\frac{1}{4}.$$

c. 6 is the geometric mean for y and x. But since $x = y + 9$, we have

$$\frac{y}{6} = \frac{6}{y + 9}$$

$$y(y + 9) = 36$$

$$y^2 + 9y - 36 = 0$$

$$(y + 12)(y - 3) = 0$$

$$y = -12 \text{ or } y = 3; \text{ reject } y = -12$$

$$\therefore y = 3 \text{ and } x = 12.$$

d. $4^2 + x^2 = \left(6\sqrt{2}\right)^2$

$$16 + x^2 = 72$$

$$x^2 = 56$$

$$x = \sqrt{56} = 2\sqrt{14} \approx 7.48$$

$$y = 13$$

41.

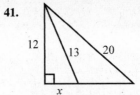

$x = 5$ using the Triple (5, 12, 13).
$y = 16$ using the Triple (12, 16, 20).
The ships are 11 km apart.

42. a. $14^2 < 12^2 + 13^2 \therefore$ acute $\triangle$

b. $11 + 5 \not> 18 \therefore$ no $\triangle$

c. $18^2 > 9^2 + 15^2 \therefore$ obtuse $\triangle$

d. $10^2 = 6^2 + 8^2 \therefore$ right $\triangle$

e. $8 + 7 \not> 16 \therefore$ no $\triangle$

f. $8^2 < 7^2 + 6^2 \therefore$ acute $\triangle$

g. $13^2 > 8^2 + 9^2 \therefore$ obtuse $\triangle$

h. $4^2 > 2^2 + 3^2 \therefore$ obtuse $\triangle$

CHAPTER TEST

1. a. $3{:}5$ or $\dfrac{3}{5}$

b. $\dfrac{25 \text{ mi}}{\text{gal}}$

2. a. $\dfrac{x}{5} = \dfrac{8}{13}$
$13x = 40$
$x = \dfrac{40}{13}$

b. $\dfrac{x+1}{5} = \dfrac{16}{x-1}$
$(x+1)(x-1) = 80$
$x^2 - 1 = 80$
$x^2 = 81$
$x = 9$ or -9

3. $15°$ and $75°$

4. a. $m\angle W \cong m\angle T$. Let $m\angle T = x$
$m\angle T + m\angle R + m\angle S = 180°$
$x + 67 + 21 = 180$
$x = 92$
$m\angle W = 92°$

b. $\dfrac{UW}{RT} = \dfrac{WV}{TS}$
$\dfrac{6}{4} = \dfrac{x}{8}$
$48 = 4x$
$x = 12$
$WV = 12$

5. a. SAS~

b. AA

6. $\triangle ABC \sim \triangle ACD \sim \triangle CBD$

7. a. $c = \sqrt{a^2 + b^2}$
$c = \sqrt{5^2 + 4^2}$
$c = \sqrt{25 + 16}$
$c = \sqrt{41}$

b. $a^2 + b^2 = c^2$
$a^2 = c^2 - b^2$
$a = \sqrt{c^2 - b^2}$
$a = \sqrt{8^2 - 6^2}$
$a = \sqrt{64 - 36}$
$a = \sqrt{28} = 2\sqrt{7}$

8. a. $a^2 + b^2 = 15^2 + 8^2 = 225 + 64 = 289$
$c^2 = 17^2 = 289$
Yes

b. $a^2 + b^2 = 11^2 + 8^2 = 121 + 64 = 185$
$c^2 = 15^2 = 225$
No

9. $(AC)^2 = (AB)^2 + (BC)^2$
$AC = \sqrt{4^2 + 3^2}$
$AC = \sqrt{16 + 9}$
$AC = \sqrt{25} = 5$
$(DA)^2 = (AC)^2 + (DC)^2$
$DA = \sqrt{5^2 + 8^2}$
$DA = \sqrt{25 + 64}$
$DA = \sqrt{89}$

10. a. In the 45-45-90 $\triangle$, $XZ = 10$ and $YZ = 10$ so
$XY = 10\sqrt{2}$ in.

b. In the 45-45-90 $\triangle$, $XY = 8\sqrt{2}$ so
$XZ = 8$ cm.

11. a. In the 30-60-90 $\triangle$, $EF = 10$ so $DE = 5$ m .

b. In the 30-60-90 $\triangle$, $DF = 6\sqrt{3}$ so $EF = 12$ ft .

12. Let $EC = x$. Then $AC = 9 + x$.
$\dfrac{AB}{AD} = \dfrac{AC}{AE}$
$\dfrac{6+8}{6} = \dfrac{9+x}{9}$
$6(9 + x) = 126$
$54 + 6x = 126$
$6x = 72$
$x = 12$
$EC = 12$

13. Let $PQ = x$. Then $QM = 10 - x$.

$$\frac{PQ}{PN} = \frac{QM}{MN}$$

$$\frac{x}{6} = \frac{10 - x}{9}$$

$$9x = 6(10 - x)$$

$$9x = 60 - 6x$$

$$15x = 60$$

$$x = 4$$

$$PQ = 4$$

$$QM = 10 - 4 = 6$$

14. 1

15. If $\angle 1 \cong \angle C$ and $\angle B \cong \angle B$,

then $\triangle ACB \sim \triangle MDB$. By CPSTP,

$$\frac{CB}{DB} = \frac{AB}{MB} \quad \text{or} \quad \frac{12}{x} = \frac{14 + x}{6}$$

$$14x + x^2 = 72$$

$$x^2 + 14x - 72 = 0$$

$$(x + 18)(x - 4) = 0$$

$$x = -18 \text{ or } x = 4; \text{ reject} = -18$$

$$\therefore DB = 4$$

16. **S1.** $\overline{MN} \parallel \overline{QR}$

R1. Given

R2. Corresponding $\angle$ s are $\cong$.

S3. $\angle P \cong \angle P$

R4. AA

17. **R1.** Given

R2. Identity

R3. Given

R5. Substitution

R6. SAS~

S7. $\angle PRC \cong \angle B$.

Chapter 6: Circles

SECTION 6.1: Circles and Related Segments and Angles

1. 29°

5. 56.6°

9. **a.** 90°

 b. 270°

 c. 135°

 d. 135°

13. **a.** 72°

 b. 144°

 c. 36°

 d. 72°

 e. Draw in $\overline{OA}$. In $\triangle BOA$, $m\angle BOA = 144°$; therefore the $m\angle ABO = 18°$

17. $RV = 4$ and let $RQ = x$.
 Using the Pythagorean Theorem, we have
 $$x^2 + 4^2 = (x+2)^2$$
 $$x^2 + 16 = x^2 + 4x + 4$$
 $$12 = 4x$$
 $$x = 3$$
 $$RQ = 3$$

21. 90°; Square

25. **a.**

 At 6:30 PM, the hour hand is half the distance from 6 to the 7. Therefore, the angle measure is 15°.

 b.

 At 5:40 AM the hour hand is $\frac{2}{3}$ the distance from 5 to 6. Therefore, the angle measure is found by adding 60 and 10. The angle is 70°.

29. 45°

33. Proof: Using the chords $\overline{AB}$, $\overline{BC}$, $\overline{CD}$ and $\overline{AD}$ in $\odot O$ as sides of inscribed angles, $\angle B \cong \angle D$ and $\angle A \cong \angle C$ since they are inscribed angles intercepting the same arc. $\triangle ABE \sim \triangle CED$ by AA.

37. Prove: If two inscribed angles intercept the same arc, then these angles are congruent.
 Given: A circle with inscribed angles A and D intercepting $\overset{\frown}{BC}$.
 Prove: $\angle A \cong \angle D$

 Proof: Since both inscribed angles, A and D, intercept $\overset{\frown}{BC}$, $m\angle A = \frac{1}{2} m\overset{\frown}{BC}$ and
 $m\angle D = \frac{1}{2} m\overset{\frown}{BC}$. Therefore, $m\angle A = m\angle D$ which means $\angle A \cong \angle D$.

41. Given: $\odot O$ with inscribed $\angle RSW$ and diameter $\overline{ST}$
 Prove: $m\angle RSW = \frac{1}{2} \cdot m\overset{\frown}{RW}$

 Proof: In $\odot O$, diameter $\overline{ST}$ is one side of inscribed $\angle RST$ and one side of inscribed $\angle TSW$. Using Case (1), $m\angle RST = \frac{1}{2} \cdot m\overset{\frown}{RT}$ and
 $m\angle TSW = \frac{1}{2} \cdot m\overset{\frown}{TW}$. By the Addition Property of Equality,
 $m\angle RST + m\angle TSW = \frac{1}{2} \cdot m\overset{\frown}{RT} + \frac{1}{2} \cdot m\overset{\frown}{TW}$. But since $m\angle RSW = m\angle RST + m\angle TSW$, we have
 $m\angle RSW = \frac{1}{2} \cdot m\overset{\frown}{RT} + \frac{1}{2} \cdot m\overset{\frown}{TW}$. Factoring the $\frac{1}{2}$ out, we have $m\angle RSW = \frac{1}{2}\left(m\overset{\frown}{RT} + m\overset{\frown}{TW}\right)$. But
 $m\overset{\frown}{RT} + m\overset{\frown}{TW} = m\overset{\frown}{RW}$. Therefore
 $m\angle RSW = \frac{1}{2} \cdot m\overset{\frown}{RW}$.

SECTION 6.2: More Angle Relationships in the Circle

1. If $m\widehat{AB} = 92°$, $m\widehat{DA} = 114°$, and $m\widehat{BC} = 138°$, then $m\widehat{DC} = 16°$.

   a. $m\angle 1 = \frac{1}{2}(16) = 8°$

   b. $m\angle 2 = \frac{1}{2}(92) = 46°$

   c. $m\angle 3 = \frac{1}{2}(92 - 16) = \frac{1}{2}(76) = 38°$

   d. $m\angle 4 = \frac{1}{2}(92 + 16) = \frac{1}{2}(108) = 54°$

   e. $m\angle 5 = 180 - 54 = 126°$ or
   $m\angle 5 = \frac{1}{2}(114 + 138) = \frac{1}{2}(252) = 126°$

5. Let $m\widehat{RT} = x$, then $m\widehat{TS} = 4x$
   $x + 4x = 180$
   $5x = 180$
   $x = 36$
   $m\widehat{RT} = 36°$
   $m\angle RST = \frac{1}{2}(m\widehat{RT})$
   $m\angle RST = \frac{1}{2}(36)$
   $m\angle RST = 18°$

9. a. $m\angle ACB = \frac{1}{2}(m\widehat{BC})$ or $68 = \frac{1}{2}(m\widehat{BC})$ or
   $m\widehat{BC} = 136°$.

   b. $m\widehat{BDC} = 360 - 136 = 224°$

   c. $m\angle ABC = \frac{1}{2}(136) = 68°$

   d. $m\angle A = \frac{1}{2}(224 - 136) = \frac{1}{2}(88) = 44°$

13. a. $m\widehat{RT} = 120°$

    b. $m\widehat{RST} = 240°$

    c. $m\angle 3 = \frac{1}{2}(240 - 120) = \frac{1}{2}(120) = 60°$

17. Let $m\widehat{CE} = x$ and $m\widehat{BD} = y$.
    $62 = \frac{1}{2}(x + y)$
    $26 = \frac{1}{2}(x - y)$

    $124 = x + y$
    $\underline{52 = x - y}$
    $176 = 2x$
    $88 = x$

    $m\widehat{CE} = 88°$
    $y = 36$
    $m\widehat{BD} = 36°$

21. 1. $\overline{AB}$ and $\overline{AC}$ are tangents to $\odot O$ from A

    2. Measure of an $\angle$ formed by a tangent and a chord $= \frac{1}{2}$ the arc measure.

    3. Substitution

    4. If 2 $\angle$ s are = in measure, they are $\cong$.

    5. $\overline{AB} \cong \overline{AC}$

    6. $\triangle ABC$ is isosceles

25. If $m\widehat{AB} = x$, then $m\widehat{ADB} = 360 - x$. Then
    $m\angle 1 = \frac{1}{2}(m\widehat{ADB} - m\widehat{AB})$
    $= \frac{1}{2}(360 - x - x)$
    $= \frac{1}{2}(360 - 2x)$
    $= 180° - x$

29. Each arc of the circle is 72°.
    $m\angle 1 = \frac{1}{2}(72) = 36°$
    $m\angle 2 = \frac{1}{2}(144 + 72) = \frac{1}{2}(216) = 108°$

33. $\angle x \cong \angle x$; $\angle R \cong \angle W$; also, $\angle RVW \cong \angle WSX$

39. If 2 parallel lines intersect a circle, then the intercepted arcs between these lines are congruent.

Given: $\overline{BC} \parallel \overline{AD}$

Prove: $\overparen{AB} \cong \overparen{CD}$

Proof: Draw $\overline{AC}$. If $\overline{BC} \parallel \overline{AD}$, then $\angle 1 \cong \angle 2$ or $m\angle 1 = m\angle 2$. $m\angle 1 = \frac{1}{2}m\overparen{AB}$ and

$m\angle 2 = \frac{1}{2}m\overparen{CD}$. Therefore, $\frac{1}{2}m\overparen{AB} = \frac{1}{2}m\overparen{CD}$ or

$m\overparen{AB} = m\overparen{CD}$. $\overparen{AB} \cong \overparen{CD}$ since they are in the same circle and have equal measures.

43. If one side of an inscribed triangle is a diameter, then the triangle is a right triangle.

Given: $\triangle ACB$ inscribed in $\odot O$ with $\overline{AB}$ a diameter

Prove: $\triangle ABC$ is a right $\triangle$

Proof: If $\triangle ACB$ is inscribed in $\odot O$ and $\overline{AB}$ is a diameter, then $\angle ACB$ must be a right angle because an angle inscribed in a semicircle is a right angle. $\triangle ACB$ is a right triangle since it contains a right angle.

47. Given: Quad. *RSTV* inscribed in $\odot Q$

Prove: $m\angle R + m\angle T = m\angle V + m\angle S$

Proof: $m\angle R = \frac{1}{2}m\overparen{VTS}$ and $m\angle T = \frac{1}{2}m\overparen{VRS}$

$\therefore m\angle R + m\angle T = \frac{1}{2}m\overparen{VTS} + \frac{1}{2}m\overparen{VRS}$

$= \frac{1}{2}\left(m\overparen{VTS} + m\overparen{VRS}\right)$

$= \frac{1}{2}(360°)$

$= 180°$

Similarly,

$m\angle V + m\angle S = \frac{1}{2}m\overparen{RST} + \frac{1}{2}m\overparen{RVT}$

$= \frac{1}{2}\left(m\overparen{RST} + m\overparen{RVT}\right)$

$= \frac{1}{2}(360°)$

$= 180°$

$\therefore m\angle R + m\angle T = m\angle V + m\angle S$

SECTION 6.3: Line and Segment Relationships in the Circle

1. $\triangle OCD$ is an equilateral triangle with $m\angle COD = 60°$. Since $\overline{OE} \perp \overline{CD}$, $\overline{OF}$ bisects $\overparen{CD}$. $m\overparen{CF} = m\overparen{FD}$ and

$m\angle COF = m\angle FOD = 30°$. $m\overparen{CF}$ must equal $30°$.

5. a.

b.

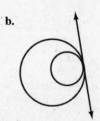

c.

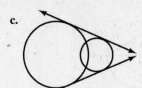

d.

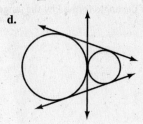

e.

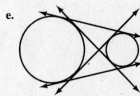

9. Let $DE = x$ and $EC = 16 - x$. Then
$$8 \cdot 6 = x(16 - x)$$
$$48 = 16x - x^2$$
$$x^2 - 16x - 48 = 0$$
$$(x - 4)(x - 12) = 0$$
$$x = 4 \text{ or } x = 12$$
$$DE = 4 \text{ and } EC = 12 \text{ OR}$$
$$DE = 12 \text{ and } EC = 4$$

15. Let $EC = x$, then $DE = 2x$.
$$9 \cdot 8 = 2x \cdot x$$
$$72 = 2x^2$$
$$36 = x^2$$
$$x^2 - 36 = 0$$
$$(x - 6)(x + 6) = 0$$
$$x = 6 \text{ or } x = -6 \text{; reject } x = -6$$
$$EC = 6$$
$$DE = 12$$

19. If $AB = 4$ and $BC = 5$, then $AC = 9$.
Let $DE = x$ and $AE = x + 3$.
$$4 \cdot 9 = 3(x + 3)$$
$$36 = 3x + 9$$
$$27 = 3x$$
$$x = 9$$
$$DE = 9$$

23. Let $RS = TV = x$ and $RV = 6 + x$.
$$x^2 = 6(6 + x)$$
$$x^2 = 36 + 6x$$
$$x^2 - 6x - 36 = 0$$
Since the quadratic won't factor, we must use the quadratic formula.
$$x = \frac{-b \pm \sqrt{b^2 - 4ac}}{2a}$$
$$x = \frac{6 \pm \sqrt{36 - 4(1)(-36)}}{2(1)}$$
$$x = \frac{6 \pm \sqrt{36 + 144}}{2}$$
$$x = \frac{6 \pm \sqrt{180}}{2}$$
$$x = \frac{6 \pm 6\sqrt{5}}{2}$$
$$x = 3 \pm 3\sqrt{5}$$
$$RS = 3 \pm 3\sqrt{5}$$

27. If $\overline{AF}$ is tangent to $\odot O$ and $\overline{AC}$ is a secant to $\odot O$, then $(AF)^2 = AC \cdot AB$. If $\overline{AF}$ is a tangent to $\odot Q$ and $\overline{AE}$ is a secant to $\odot Q$, then $(AF)^2 = AE \cdot AD$. By substitution, $AC \cdot AB = AE \cdot AD$.

31. Yes; $\overline{AE} \cong \overline{CE}$; $\overline{DE} \cong \overline{EB}$

35. Let $AM = x$, then $MB = 14 - x$.
Let $BN = y$, then $NC = 16 - y$.
Let $PC = z$, then $AP = 12 - z$.
If tangent segments to a circle from an external point are congruent, $AM = AP$, $BN = MB$, and $PC = NC$ or
$$\begin{cases} x = 12 - z \\ y = 14 - x \\ z = 16 - y \end{cases} \text{ or } \begin{cases} x + z = 12 \\ x + y = 14 \\ y + z = 16 \end{cases}$$
Subtracting the first equation from the 2nd equation and using the 3rd equation, we have
$$y - z = 2$$
$$y + z = 16$$
Adding gives $2y = 18$ or $y = 9$. Solving for x and z we have $x = 5$ and $z = 7$. Therefore, $AM = 5$; $PC = 7$ and $BN = 9$.

39. $OA = 2$; $BP = 3$; $OD = x$; $DP = 10 - x$.
$$\frac{2}{x} = \frac{3}{10 - x} \text{ or } 20 - 2x = 3x. \text{ Solve to get } x = 4.$$
$\therefore OD = 4$; $DP = 6$; $AD = 2\sqrt{3}$ and $DB = 3\sqrt{3}$ so $AB = 5\sqrt{3}$ or $AB \approx 8.7$ in.

43. Let x represent the angle measure of the larger gear. It is intuitively obvious that the

$$\frac{\text{number of teeth in the larger gear}}{\text{number of teeth in the smaller gear}} =$$

$$\frac{\text{angle measure in smaller gear}}{\text{angle measure in larger gear}}.$$

The proportion becomes

$$\frac{2}{1} = \frac{90}{x}$$
$$2x = 90$$
$$x = 45°$$

47. Given: $\overline{TX}$ is a secant segment and
$\overline{TV}$ is a tangent at V

Prove: $(TV)^2 = TW \cdot TX$

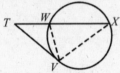

Proof: With secant $\overline{TX}$ and tangent $\overline{TV}$, draw in $\overline{WV}$ and $\overline{VX}$. $m\angle X = \frac{1}{2}m\widehat{WV}$ since $\angle X$ is an inscribed angle. $m\angle TVW = \frac{1}{2}m\widehat{WV}$ because it is formed by a tangent and a chord. By substitution, $m\angle TVW = m\angle X$ or $\angle TVW \cong \angle X$. $\angle T \cong \angle T$ and $\triangle TVW \sim \triangle TXV$. It follows that $\frac{TV}{TW} = \frac{TX}{TV}$ or $(TV)^2 = TW \cdot TX$.

SECTION 6.4: Some Constructions and Inequalities for the Circle

1. $m\angle CQD < m\angle AQB$

5. $CD < AB$

9.

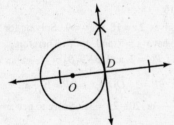

13.

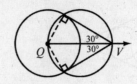

The measure of the angle formed by the tangents at V is $60°$.

17. a. $\overline{OT}$

   b. $\overline{OD}$

21. Obtuse

25. a. $m\angle\widehat{AB} > m\angle\widehat{BC}$

   b. $AB > BC$

29. a. $\angle B$

   b. $\overline{AC}$

33.

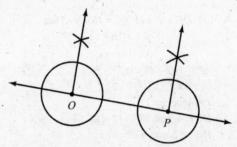

37. Draw in $\overline{OB}$ to form rt. $\triangle OMB$. Since $MB = 12$ and $OB = 13$, $OM = 5$. Draw in $\overline{OD}$ to form rt. $\triangle OND$. Since $ND = 5$ and $OD = 13$, $ON = 12$.

CHAPTER REVIEW

1. (9, 12, 15) is a multiple of the Pythagorean Triple (3, 4, 5). Therefore, the distance from the center of the circle to the chord is 9 mm.

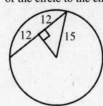

2. (8, 15, 17) is a Pythagorean Triple. Therefore, the length of half of the chord is 15 and the length of the chord is 30 cm.

3. $r^2 = 5^2 + 4^2$
$r^2 = 25 + 16$
$r^2 = 41$
$r = \sqrt{41}$

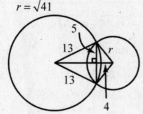

4. The radius of each circle has a length of $6\sqrt{2}$.

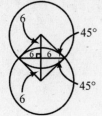

5. $m\angle B = \frac{1}{2}\left(m\widehat{AD} - m\widehat{AC}\right)$
$25 = \frac{1}{2}\left(140 - m\widehat{AC}\right)$
$50 = 140 - m\widehat{AC}$
$m\widehat{AC} = 90°$
$m\widehat{DC} = 360 - (140 + 90) = 130°$

6. $m\widehat{AC} = 360 - (140 + 155) = 65°$
$m\angle B = \frac{1}{2}\left(m\widehat{AD} - m\widehat{AC}\right)$
$m\angle B = \frac{1}{2}(155 - 65) = \frac{1}{2}(90) = 45°$

7. If $m\angle EAD = 70°$ then $m\widehat{AD} = 140°$.
$m\angle B = \frac{1}{2}\left(m\widehat{AD} - m\widehat{AC}\right)$
$30 = \frac{1}{2}\left(140 - m\widehat{AC}\right)$
$60 = 140 - m\widehat{AC}$
$m\widehat{AC} = 80°$

8. If $m\angle D = 40°$ then $m\widehat{AC} = 80°$.
$m\widehat{AD} = 360 - (80 + 130) = 150°$
$m\angle B = \frac{1}{2}\left(m\widehat{AD} - m\widehat{AC}\right)$
$m\angle B = \frac{1}{2}(150 - 80)$
$m\angle B = \frac{1}{2}(70) = 35°$

9. Let $m\widehat{AC} = m\widehat{CD} = x$. Then $m\widehat{AD} = 360 - 2x$.
$m\angle B = \frac{1}{2}\left(m\widehat{AD} - m\widehat{AC}\right)$
$40 = \frac{1}{2}(360 - 2x - x)$
$80 = 360 - 3x$
$-280 = -3x$
$x = 93\frac{1}{3}$
$m\widehat{AC} = m\widehat{DC} = 93\frac{1}{3}°$
$m\widehat{AD} = 173\frac{1}{3}°$

10. Let $m\widehat{AC} = x$; $m\widehat{AD} = 290 - x$
$m\angle B = \frac{1}{2}\left(m\widehat{AD} - m\widehat{AC}\right)$
$35 = \frac{1}{2}(290 - x - x)$
$70 = 290 - 2x$
$-220 = -2x$
$x = 110$
$m\widehat{AC} = 110°$ and $m\widehat{AD} = 80°$

11. If $m\angle 1 = 46°$, then $m\widehat{BC} = 92°$.
If $\overline{AC}$ is a diameter, $m\widehat{AB} = 88°$.
$m\angle 2 = 44°$; $m\angle 3 = 90°$; $m\angle 4 = 46°$;
$m\angle 5 = 44°$.

12. If $m\angle 5 = 40°$, then $m\widehat{AB} = 80°$.
If $\overline{AC}$ is a diameter, $m\widehat{BC} = 100°$.
$m\angle 1 = 50°$; $m\angle 2 = 40°$; $m\angle 3 = 90°$;
$m\angle 4 = 50°$.

13. (12, 16, 20) is a multiple of the Pythagorean Triple (3, 4, 5). Hence, half of the chord has a length of 12 and the chord has length 24.

14. The radius of the circle has length 10 using the Pythagorean Triple (6, 8, 10).

15. A

16. S

17. N

18. S

19. A

20. N

21. A

22. N

23. a. $m\angle AEB = \frac{1}{2}(m\widehat{AB} + m\widehat{CD})$
$75 = \frac{1}{2}(80 + m\widehat{CD})$
$150 = 80 + m\widehat{CD}$
$m\widehat{CD} = 70°$

b. $m\angle BED = \frac{1}{2}(m\widehat{AC} + m\widehat{BD})$
$45 = \frac{1}{2}(62 + m\widehat{BD})$
$90 = 62 + m\widehat{BD}$
$m\widehat{BD} = 28°$

c. $m\angle P = \frac{1}{2}(m\widehat{AB} - m\widehat{CD})$
$24 = \frac{1}{2}(88 - m\widehat{CD})$
$48 = 88 + m\widehat{CD}$
$m\widehat{CD} = 40°$
$m\angle CED = \frac{1}{2}(m\widehat{AB} + m\widehat{CD})$
$m\angle CED = \frac{1}{2}(88 + 40) = \frac{1}{2}(128) = 64°$

d. $m\angle CED = \frac{1}{2}(m\widehat{AB} + m\widehat{CD})$
$41 = \frac{1}{2}(m\widehat{AB} + 20)$
$82 = m\widehat{AB} + 20$
$m\widehat{AB} = 62°$
$m\angle P = \frac{1}{2}(m\widehat{AB} - m\widehat{CD})$
$m\angle P = \frac{1}{2}(62 - 20) = \frac{1}{2}(42) = 21°$

e. $m\angle AEB = \frac{1}{2}(m\widehat{AB} + m\widehat{CD})$ and
$m\angle P = \frac{1}{2}(m\widehat{AB} - m\widehat{CD})$.

$65 = \frac{1}{2}(m\widehat{AB} + m\widehat{CD})$
$25 = \frac{1}{2}(m\widehat{AB} - m\widehat{CD})$

$130 = m\widehat{AB} + m\widehat{CD}$
$\underline{50 = m\widehat{AB} - m\widehat{CD}}$
$180 = 2 \cdot m\widehat{AB}$
$m\widehat{AB} = 90°$; $m\widehat{CD} = 90°$

23. f. $m\angle CED = \frac{1}{2}(m\widehat{AB} - m\widehat{CD})$
$50 = \frac{1}{2}(m\widehat{AB} - m\widehat{CD})$
$100 = (m\widehat{AB} - m\widehat{CD})$
$m\widehat{AC} + m\widehat{BD} = 360 - 100 = 260°$

24. a. Let $BC = x$.
$6^2 = 12 \cdot x$
$36 = 12x$
$x = 3$
$BC = 3$

b. Let $DG = x$.
$4 \cdot 6 = x \cdot 3$
$24 = 3x$
$x = 8$
$DG = 8$

c. Let $CE = x$.
$3 \cdot x = 4 \cdot 12$
$3x = 48$
$x = 16$
$CE = 16$

d. Let $GE = x$.
$10 \cdot x = 5 \cdot 8$
$10x = 40$
$x = 4$
$GE = 4$

e. Let $BC = x$ and $CA = x + 5$.
$6^2 = x(x + 5)$
$36 = x^2 + 5x$
$0 = x^2 + 5x - 36$
$0 = (x + 9)(x - 4)$
$x = -9$ or $x = 4$; reject $x = -9$.
$BC = 4$.

f. Let $DG = x$ and $AG = 9 - x$.

$$x(9 - x) = 4 \cdot 2$$
$$9x - x^2 = 8$$
$$0 = x^2 - 9x + 8$$
$$0 = (x - 8)(x - 1)$$
$$x = 8 \text{ or } x = 1;$$
$$GD = 8 \text{ or } GD = 1.$$

g. Let $CD = ED = x$ and $CE = 2x$.

$$x(2x) = 3 \cdot 30$$
$$2x^2 = 90$$
$$x^2 = 45$$
$$x = \sqrt{45} = 3\sqrt{5}$$
$$ED = 3\sqrt{5}$$

h. Let $CD = x$ and $CE = x + 12$

$$x(x + 12) = 5 \cdot 9$$
$$x^2 + 12x = 45$$
$$x^2 + 12x - 45 = 0$$
$$(x + 15)(x - 3) = 0$$
$$x = -15 \text{ or } x = 3; \text{ reject } x = -15;$$
$$CD = 3.$$

i. Let $FC = x$

$$x^2 = 4 \cdot 12$$
$$x^2 = 48$$
$$x = \sqrt{48} = 4\sqrt{3}$$
$$FC = 4\sqrt{3}$$

j. Let $CD = x$ and $CE = x + 9$.

$$x(x + 9) = 6^2$$
$$x^2 + 9x = 36$$
$$x^2 + 9x - 36 = 0$$
$$(x + 12)(x - 3) = 0$$
$$x = -12 \text{ or } x = 3; \text{ reject } x = -12;$$
$$CD = 3$$

25. $5x + 4 = 2x + 19$
$$3x = 15$$
$$x = 5$$
$$OE = 5(5) + 4 = 29$$

26. $x(x - 2) = x + 28$
$$x^2 - 2x - x - 28 = 0$$
$$x^2 - 3x - 28 = 0$$
$$(x - 7)(x + 4) = 0$$
$x = 7$ or $x = -4$. If $x = 7$, then
$AC = 7 + 28 = 35$; $DE = 17\frac{1}{2}$.
If $x = -4$, then $AC = 24$; $DE = 12$.

27. Given: $\overline{DC}$ is tangent to circles B and A at points D and C, respectively.
Prove: $AC \cdot ED = CE \cdot BD$

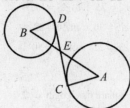

Proof: If $\overline{DC}$ is tangent to circles B and A at points D and C, then $\overline{BD} \perp \overline{DC}$ and $\overline{AC} \perp \overline{DC}$. $\angle$s D and C are congruent since they are right angles. $\angle DEB \cong \angle CEA$ because of vertical angles. $\triangle BDE \sim \triangle ACE$ by AA. It follows that $\dfrac{AC}{CE} = \dfrac{BD}{ED}$ since corresponding sides are proportional. Hence, $AC \cdot ED = CE \cdot BD$.

28. Given: $\odot O$ with $\overline{EO} \perp \overline{BC}$,
$\overline{DO} \perp \overline{BA}$, $\overline{EO} \cong \overline{OD}$
Prove: $\overparen{BC} \cong \overparen{BA}$

Proof: In $\odot O$, if $\overline{EO} \perp \overline{BC}$, $\overline{DO} \perp \overline{BA}$, and $\overline{EO} \cong \overline{OD}$, then $\overline{BC} \cong \overline{BA}$. (Chords equidistant from the center of the circle are congruent.) It follows then that $\overparen{BC} \cong \overparen{BA}$.

29. Given: $\overline{AP}$ and $\overline{BP}$ are tangent to $\odot Q$ at A and B; C is the midpoint of $\overparen{AB}$.
Prove: $\overrightarrow{PC}$ bisects $\angle APB$

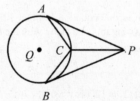

Proof: If $\overline{AP}$ and $\overline{BP}$ are tangent to $\odot Q$ at A and B, then $\overline{AP} \cong \overline{BP}$. $\overparen{AC} \cong \overparen{BC}$ since C is the midpoint of $\overparen{AB}$. It follows that $\overline{AC} \cong \overline{BC}$ and using $\overline{CP} \cong \overline{CP}$, we have $\triangle ACP \cong \triangle BCP$ by SSS. $\angle APC \cong \angle BCP$ by CPCTC and hence $\overrightarrow{PC}$ bisects $\angle APB$.

30. If $m\overset{\frown}{AD} = 136°$ and $\overline{AC}$ is a diameter, then

$m\overset{\frown}{DC} = 44°$. If $m\overset{\frown}{BC} = 50°$, then $m\overset{\frown}{AB} = 130°$.

$m\angle 1 = 93°$; $m\angle 2 = 25°$; $m\angle 3 = 43°$;

$m\angle 4 = 68°$; $m\angle 5 = 90°$; $m\angle 6 = 22°$;

$m\angle 7 = 68°$; $m\angle 8 = 22°$; $m\angle 9 = 50°$;

$m\angle 10 = 112°$

31. Each side of the square has length $6\sqrt{2}$.

Therefore, the perimeter is $24\sqrt{2}$ cm.

32. The perimeter of the triangle is $15 + 5\sqrt{3}$ cm.

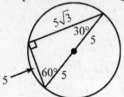

33.
$$(35 - x)^2 + (6 + x)^2 = 29^2$$
$$1225 - 70x + x^2 + 36 + 12x + x^2 = 841$$
$$2x^2 - 58x + 420 = 0$$
$$x^2 - 29x + 210 = 0$$
$$(x - 15)(x - 14) = 0$$

$x = 15$ or $x = 14$

The lengths of the segments on the hypotenuse are 14 and 15.

34. Let $AD = x = AF$; $BE = y = DB$;

$FC = z = CE$, then

$x + y = 9$

$x + z = 10$

$ y + z = 13$

Subtracting the second equation from the first we get $y - z = 1$. Using this one along with the third equation, we have

$y - z = -1$

$y + z = 13$.

Adding, we get $2y = 12$ or $y = 6$.

Solving for x and z, $x = 3$ and $z = 7$.

$AD = 3$; $BE = 6$; $FC = 7$.

35. a. $AB > CD$

    b. $QP < QR$

    c. $m\angle A < m\angle C$

36. a.

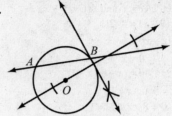

    b.

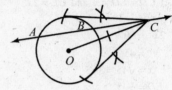

37.

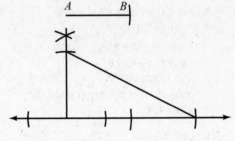

38.

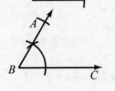

CHAPTER TEST

1. a. $272°$

  b. $m\widehat{ACB} = 360 - m\widehat{AB} = 360 - 92 = 268°$
  $m\widehat{AC} = \frac{1}{2}m\widehat{ACB} = \frac{1}{2}(268) = 134°$

2. a. $69°$

  b. $m\angle BAC = \frac{1}{2}m\widehat{BC} = \frac{1}{2}(64) = 32°$

3. a. $m\angle BAC = \frac{1}{2}m\widehat{BC}$
  $m\widehat{BC} = 2(24) = 48°$

  b. Isosceles

4. a. Right

  b. Congruent

5. a. $m\angle 1 = \frac{1}{2}(106 + 32) = \frac{1}{2}(138) = 69°$

  b. $m\angle 2 = \frac{1}{2}(106 - 32) = \frac{1}{2}(74) = 37°$

6. a. $214°$

  b. $m\angle 3 = \frac{1}{2}(214 - 146) = \frac{1}{2}(68) = 34°$

7. $m\angle 3 = \frac{1}{2}\left(m\widehat{RST} - m\widehat{RT}\right)$

  $46 = \frac{1}{2}\left(m\widehat{RST} - m\widehat{RT}\right)$

  $92 = m\widehat{RST} - m\widehat{RT}$

  $360 = m\widehat{RST} + m\widehat{RT}$

  Adding the above 2 equations gives

  $452 = 2 \cdot m\widehat{RST}$ or $m\widehat{RST} = 226$

  a. $m\widehat{RST} = 226°$ **b.** $m\widehat{RT} = 134°$

8. a. Concentric

b. $(QV)^2 = (QR)^2 + (RV)^2$
  $(RV)^2 = (QV)^2 - (QR)^2$
  $(RV)^2 = 5^2 - 3^2$
  $(RV)^2 = 25 - 9$
  $(RV)^2 = 16$
  $RV = 4$
  $TV = 8$

9. $AB = 6$; $OC = AO = 5$; $AC = 10$

  In rt. $\triangle ABC$, $BC = 8$. If M is the midpoint, then

  $MB = MC = 4$. In rt. $\triangle MBA$,

  $(AM)^2 = 6^2 + 4^2$

  $(AM)^2 = 36 + 16 = 52$

  $AM = \sqrt{52} = \sqrt{4 \cdot 13} = 2\sqrt{13}$

10. a. 1

  b. 2

11. a. $HP = 4$, $PJ = 5$, and $PM = 2$.
  Let $LP = x$.
  $4 \cdot 5 = 2 \cdot x$
  $20 = 2x$
  $x = 10$
  $LP = 10$

  b. $HP = x + 1$, $PJ = x - 1$, $LP = 8$, and
  $PM = 3$.
  $(x + 1)(x - 1) = 8 \cdot 3$
  $x^2 - 1 = 24$
  $x^2 - 25 = 0$
  $(x - 5)(x + 5) = 0$
  $x = 5$ or $x = -5$; reject $x = -5$.

12. If $TX = 3$ and $XW = 5$, then $TW = 8$.

  If $\triangle TVW \sim \triangle TXV$, then

  $\dfrac{TV}{TX} = \dfrac{TW}{TV}$ or $\dfrac{TV}{3} = \dfrac{8}{TV}$ or

  $(TV)^2 = 24$

  $TV = \sqrt{24} = \sqrt{4 \cdot 6} = 2\sqrt{6}$

13.

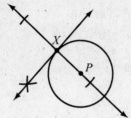

14. **a.** m∠AQB > m∠CQD

    b. *AB > CD*

15. **a.** 1

    b. 7

16. **S1.** In ⊙*O*, chords $\overline{AD}$ and $\overline{BC}$ intersect at *E*

    R1. Given

    R2. Vertical angles are congruent

    R4. AA

    S5. $\dfrac{AE}{CE} = \dfrac{BE}{DE}$

Chapter 7: Locus and Concurrence

SECTION 7.1: Locus of Points

1. A, C, E

5.

9.
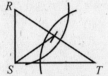

13. The locus of points at a distance of 3 inches from point O is a circle.

17. The locus of the midpoints of the chords in ⊙Q parallel to diameter $\overline{PR}$ is the perpendicular bisector of $\overline{PR}$.

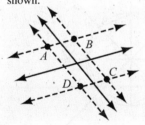

21. The locus of points 1 cm from line *p* and 2 cm from line *q* are the four points *A*, *B*, *C*, and *D* as shown.

25. The locus of points at a distance of 2 cm from a sphere whose radius is 5 cm is two concentric spheres with the same center. The radius of one sphere is 3 cm and the radius of the other sphere is 7 cm.

29. The locus of points equidistant from an 8 foot ceiling and the floor is a parallel plane in the middle.

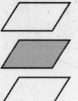

33.
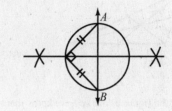

37.
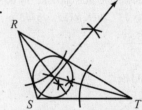

41.
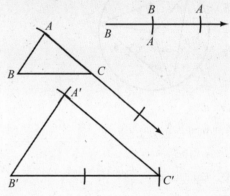

SECTION 7.2: Concurrence of Lines

1. Yes

5. Circumcenter

9. No

13. Midpoint of the hypotenuse

17.

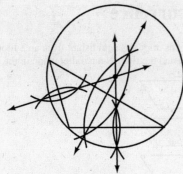

21.

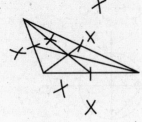

25. Using the figure as marked, we know that in the 30°-60°-90° triangle, the side opposite the 60 degree angle is $x\sqrt{3}$. Therefore, $x\sqrt{3} = 5$ or

$$x = \frac{5}{\sqrt{3}} \cdot \frac{\sqrt{3}}{\sqrt{3}} = \frac{5\sqrt{3}}{3}.$$

The radius would be $2 \cdot \frac{5\sqrt{3}}{3}$ or $\frac{10\sqrt{3}}{3}$.

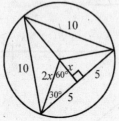

29. a. 4

   b. 6

   c. 10.5

33. Equilateral

37. a. Yes

   b. No

41. Draw isosceles $\triangle ABC$ with $AB = BC = 10$. Base $AC = 12$. Let circle O be the inscribed circle. Draw the altitude $\overline{BF} \perp \overline{AC}$. Because the $\triangle$ is isosceles $\overline{BF}$ will contain point O.

$BF = 8$. Let $OF = r$ and $OB = 8 \cdot r$. If a circle is inscribed in a $\triangle$, then the center of the circle is equidistant from the sides of the $\triangle$. If a point is equidistant from the sides of an angle, then it is on the angle bisector. $\therefore \overrightarrow{AO}$ bisects $\angle$ BAF in $\triangle$ BAF and we can use the proportion

$$\frac{AB}{AF} = \frac{OB}{OF}$$

$$\frac{10}{6} = \frac{8-r}{r}$$

$$\frac{5}{3} = \frac{8-r}{r}$$

$$5r = 24 - 3r$$

$$8r = 24$$

$$r = 3; \text{ radius } = 3 \text{ in.}$$

SECTION 7.3: More About Regular Polygons

1. First, construct the angle-bisectors of two consecutive angles, say A and B. The point of intersection, O, is the center of the inscribed circle.

Second, construct the line segment $\overline{OM}$ which is perpendicular to $\overline{AB}$. Then, using the radius $r = OM$, construct the inscribed circle with the center O.

5. In $\odot O$, draw diameter $\overline{AB}$.

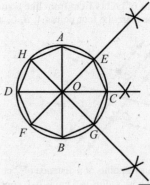

Now construct the diameter $\overline{CD}$ which is $\perp$ to $\overline{AB}$ at O. Construct the angle bisectors for $\angle$s AOC and BOC, and extend these to form diameters $\overline{EF}$ and $\overline{GH}$. Joining in the order $A, E, C, G, B, F, D,$ and H, determines a regular octagon inscribed in $\odot O$.

9. $P = 8(3.4) = 27.2$ in.

13. In square *ABCD*, apothem *a* = 5 in.

(using the 45°-45°-90° relationship).

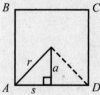

Also radius $r = 5\sqrt{2}$ in.

17. $c = \dfrac{360}{n}$

  a. 120°

  b. 90°

  c. 72°

  d. 60°

21. $c = \dfrac{360}{n} \Rightarrow n = \dfrac{360}{c}$; ext. $\angle = \dfrac{360}{n}$

  a. n = 9; ext. $\angle$ = 40; int. $\angle$ = 140°

  b. n = 8; ext. $\angle$ = 45; int. $\angle$ = 135°

  c. n = 6; ext. $\angle$ = 60; int. $\angle$ = 120°

  d. n = 4; ext. $\angle$ = 90; int. $\angle$ = 90°

25. $c = \dfrac{360}{n}$

  a. n = 9; yes

  b. n ≠ an integer; no

  c. n = 6; yes

  d. n ≠ an integer; no

29.

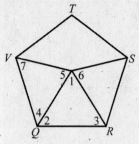

m∠1 = m∠2 = m∠3 = 60°

$m\angle VQR = \dfrac{(5-2)180}{5} = 108°$;

So m∠4 = 108 − 60 = 48°.

With $\overline{VQ} \cong \overline{QR}$ and $\overline{PQ} \cong \overline{QR}$, $\overline{VQ} \cong \overline{PQ}$,

m∠7 = m∠5

∴ m∠5 = 66°

Similarly, m∠6 = 66°.

m∠VPS = 360 − (m∠5 + m∠1 + m∠6)

m∠VPS = 168°

CHAPTER REVIEW

1.

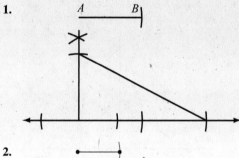

2.

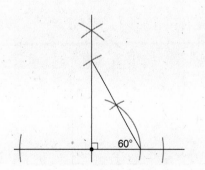

3.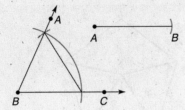

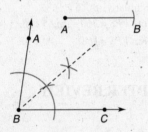

4.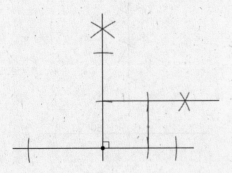

5.

6.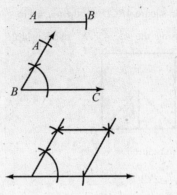

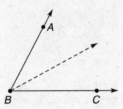

7. The locus of points equidistant from the sides of ∠ ABC is the bisector of ∠ ABC.

8. The locus of points that are 1 inch from point B is a circle with center B and radius length 1 inch.

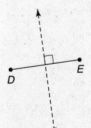

9. The locus of points equidistant from points D and E is the perpendicular bisector of $\overline{DE}$.

10. The locus of points $\frac{1}{2}$ inch from $\overleftrightarrow{DE}$ are 2

parallel lines on either side of $\overleftrightarrow{DE}$ $\frac{1}{2}$ inch from

$\overleftrightarrow{DE}$.

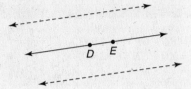

11. The locus of the midpoints of the radii of a circle is a concentric circle with radius half the length of the given radius.

12. The locus of the centers of all circles passing through two given points is the perpendicular bisector of the segment joining the 2 given points.

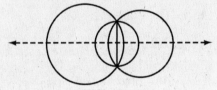

13. The locus of the centers of a penny that rolls around a half-dollar is a circle.

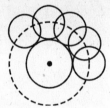

14. The locus of points in space 2 cm from point A is a sphere with center A and radius of length 2 cm.

15. The locus of points 1 cm from a given plane is 2 parallel planes on either side of the given plane at a distance of 1 cm.

16. The locus of points in space less than three units from a given point is the interior of a sphere.

17. The locus of points equidistant from two parallel planes is a parallel plane midway between the two planes.

18.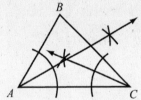

19.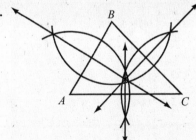

20.

21.

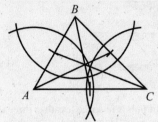

22.

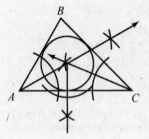

23.

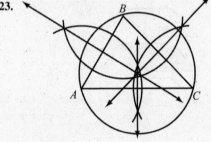

24. a. $BG = \frac{2}{3}(BF) = \frac{2}{3}(18) = 12$

    b. $AG = \frac{2}{3}(AE)$

         $4 = \frac{2}{3}(AE)$

         $AE = \frac{3}{2}(4) = 6$

         $GE = 2$

    c. $CG = \frac{2}{3}(DC)$

         $4\sqrt{3} = \frac{2}{3}(DC)$

         $DC = \frac{3}{2}(4\sqrt{3}) = 6\sqrt{3}$

         $DG = 2\sqrt{3}$

25. $AG = 2(GE)$ and $BG = 2(GF)$

     $2x + 2y = 2(2x - y)$ and $3y + 1 = 2(x)$.

     Simplifying: $2x + 2y = 4x - 2y$ and $3y + 1 = 2x$.

     $-2x + 4y = 0$ and $2x - 3y = 1$.

     Adding the above two equations gives $y = 1$.

     Solving for x gives

     $-2x + 4 = 0$

         $-2x = -4$

           $x = 2$

     $BF = BG + GF = 4 + 2 = 6$;

     $AE = AG + GE = 6 + 3 = 9$.

26. a. $c = \frac{360}{5} = 72°$

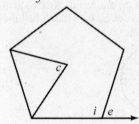

    b. $i = \frac{(5-2)180}{5} = \frac{3(180)}{5} = \frac{540}{5} = 108°$

    c. $e = 180° - 108° = 72°$

27. a. $c = \frac{360}{10} = 36°$

    b. $i = \frac{(10-2)180}{10} = \frac{8(180)}{10} = \frac{1440}{10} = 144°$

    c. $e = 180° - 144° = 36°$

28. a. $c = \frac{360}{n}$

         $45 = \frac{360}{n}$

         $45n = 360$

           $n = 8$

    b. $P = n(s)$

         $P = 8(5)$

         $P = 40$

29. a. The radius of the regular polygon is $3\sqrt{2}$. A central angle must measure 90. Using

     $c = \frac{360}{n}$, n = 4. ∴ the perimeter = 24 in.

    b. The radius length is $3\sqrt{2}$ in.

30. a. No

    b. No

    c. Yes

    d. Yes

31. a. No.

    b. Yes

    c. No

    d. Yes

32. If the radius of the inscribed circle is 7 in., then the length of the radius of the triangle is 14 in.

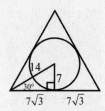

33. Draw the circle inscribed in the regular hexagon. Now draw in 2 consecutive radii of the hexagon forming an equilateral triangle. Now draw the height of the equilateral triangle (which is a radius of the circle). In the 30-60-90 triangle, the side opposite the 30°angle is a, the side opposite the 60°angle is $a\sqrt{3}$. $10 = a\sqrt{3} \Rightarrow$

$a = \dfrac{10}{\sqrt{3}} = \dfrac{10\sqrt{3}}{3}$. The side of the equilateral

$\Delta = \dfrac{20\sqrt{3}}{3}$. The perimeter of the hexagon

$= 6 \cdot \dfrac{20\sqrt{3}}{3} = 40\sqrt{3}$ cm.

CHAPTER TEST

1. The locus of points equidistant from parallel lines ℓ and m is a line parallel to ℓ and m and midway between them.

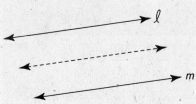

2. The locus of points equidistant from the sides of $\angle$ ABC is the angle bisector of $\angle$ ABC.

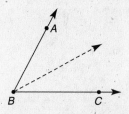

3. The locus of points equidistant from the endpoints of $\overline{DE}$ is the perpendicular bisector of $\overline{DE}$.

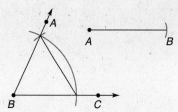

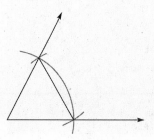

4. The locus of points in a plane that are at a distance of 3 cm from P is a circle with center P and radius of length 3 cm.

5. The locus of points in space that are at a distance of 3 cm from P is a sphere with center P and radius of length 3 cm.

6. a. Incenter   **b.** Centroid

7. a. Circumcenter   **b.** Orthocenter

8. Equilateral Δ

9. Angle bisectors and medians

10. **a.** True   **b.** True

    c. False   **d.** False

11. **a.** 1.5 in.   **b.** $3\sqrt{3}$ in.

12. **a.** $c = \dfrac{360}{n} = \dfrac{360}{5} = 72°$

    b. $i = \dfrac{(n-2)\cdot 180}{n} = \dfrac{(5-2)\cdot 180}{5} = 108°$

13. $c = \dfrac{360}{n}$

    $36 = \dfrac{360}{n}$

    $n = 10$

14. Side $= 10$ $\therefore$ P $= 80$ cm

15. **a.** $4\sqrt{3}$ in.

    b. 8 in.

Chapter 8: Areas of Polygons and Circles

SECTION 8.1: Area and Initial Postulates

1. Two triangles with equal areas are not necessarily congruent.

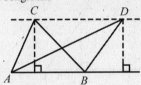

$\triangle ABC \ncong \triangle ABD$

Two squares with equal areas must be congruent because the sides will be congruent.

5. The altitudes to $\overline{PN}$ and to $\overline{MN}$ are congruent.

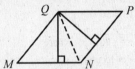

This follows from the fact that $\triangle$s QMN and QPN are congruent. Corresponding altitudes of $\cong$ $\triangle$s are $\cong$.

9. $A = bh$
$A = 6 \cdot 9$
$A = 54 \text{ cm}^2$

13. $A = bh$
$A = 12 \cdot 6$
$A = 72 \text{ in}^2$

17. Using the Pythagorean Triple, (5, 12, 13), $b = 12$.

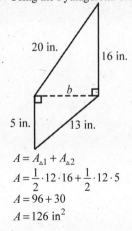

$A = A_{\triangle 1} + A_{\triangle 2}$
$A = \frac{1}{2} \cdot 12 \cdot 16 + \frac{1}{2} \cdot 12 \cdot 5$
$A = 96 + 30$
$A = 126 \text{ in}^2$

21. In the right $\triangle$, $h = 8$, using the Pythagorean Triple (6, 8, 10).

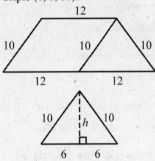

$A = A_{\text{PARALLELOGRAM}} + A_{\triangle}$
$A = 12 \cdot 8 + \frac{1}{2} \cdot 12 \cdot 8$
$A = 96 + 48$
$A = 144 \text{ units}^2$

25. $A = \frac{1}{2}bh + bh$

a. $A = \frac{1}{2} \cdot 24 \cdot 5 + 24 \cdot 10$
$A = 300 \text{ ft}^2$

b. 3 gallons

c. \$46.50

29. a. 3 ft = 1 yd $\therefore$ 9 sq ft = 1 sq yd

b. 36 in = 1 yd $\therefore$ 1296 sq in = 1 sq yd

33. Given: $\triangle ABC$ with midpoints M, N, and P
Explain why $A_{ABC} = 4 \cdot A_{MNP}$

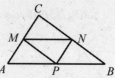

Explanation: $\overline{MN}$ joins the midpoints of $\overline{CA}$ and $\overline{CB}$ so $MN = \frac{1}{2}(AB)$. $\therefore \overline{AP} \cong \overline{PB} \cong \overline{MN}$.

$\overline{PN}$ joins the midpoints of $\overline{CB}$ and $\overline{AB}$ so $PN = \frac{1}{2}(AC)$. $\therefore \overline{AM} \cong \overline{MC} \cong \overline{PN}$.

$\overline{MP}$ joins the midpoints of $\overline{AB}$ and $\overline{AC}$ so $MP = \frac{1}{2}(BC)$. $\therefore \overline{CN} \cong \overline{NB} \cong \overline{MP}$. The four

triangles are all $\cong$ by SSS. Therefore, the area of each triangle is the same. Hence the area of the big triangle is the same as four times the area of one of the smaller triangles.

37. $A = \frac{1}{2}bh$

$40 = \frac{1}{2} \cdot x(x+2)$

Multiplying by 2,
$80 = x(x+2)$
$80 = x^2 + 2x$
$0 = x^2 + 2x - 80$
$0 = (x+10)(x-8)$
$x+10 = 0 \quad$ or $\quad x - 8 = 0$
$\quad\quad x = -10 \quad$ or $\quad\quad x = 8$

Reject $x = -10$.

41.
$A_{\text{RECT.}} = bh$
$A_{\text{LARGE RECT.}} = (b+0.2b)(h+0.3h)$
$\quad\quad\quad\quad = (1.2b)(1.3h) = 1.56bh$
$A_{\text{LARGE RECT.}} - A_{\text{RECT.}} = 1.56bh - bh = 0.56bh$
The area is increased by 56%.

45. $(a+b)(c+d) = ac + ad + bc + bd$

49. $A_{MNPQ} = (QR)(MN)$

But $MN = QP = 12$

$\therefore A_{MNPQ} = 6(12) = 72$ units^2

$A_{MNPQ} = (QS)(PN)$

But $PN = QM = 9$ so
$A_{MNPQ} = (QS)9$
$\quad\quad 72 = 9(QS)$
$\quad\quad QS = 8$ units

SECTION 8.2: Perimeter and Area of Polygons

1. $c = 13$ using the Pythagorean Triple (5, 12, 13).
$P = a + b + c$
$P = 5 + 12 + 13$
$P = 30$ in.

5.

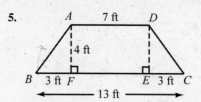

Draw the altitude from D to $\overline{BC}$. $FE = 7$ and $BF = ED = 3$. $AB = DC = 5$. Therefore, the perimeter of isosceles trapezoid $ADCB$ with altitudes $\overline{AF}$ and $\overline{DE}$
$P = 7 + 5 + 5 + 13 = 30$ ft.

9. $A = \sqrt{s(s-a)(s-b)(s-c)}$ where
$a = 13$, $b = 14$, $c = 15$ and
$s = \frac{1}{2}(13+14+15)$
$s = \frac{1}{2}(42) = 21$ $\therefore$

$A = \sqrt{21(21-13)(21-14)(21-15)}$
$A = \sqrt{21(8)(7)(6)}$
$A = \sqrt{(3 \cdot 7) \cdot (2 \cdot 4) \cdot 7 \cdot (2 \cdot 3)}$
$A = \sqrt{2^2 \cdot 3^2 \cdot 2^2 \cdot 7^2}$
$A = 2 \cdot 3 \cdot 2 \cdot 7$
$A = 84$ in^2

13. $A = \frac{1}{2}h(b_1 + b_2)$
$A = \frac{1}{2} \cdot 4(7+13)$
$A = \frac{1}{2} \cdot 4(20)$
$A = 40$ ft^2

17.

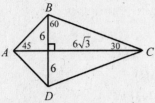

$$A = \frac{1}{2} \cdot d_1 \cdot d_2$$

$$A = \frac{1}{2} \cdot 12(6 + 6\sqrt{3})$$

$$A = 6(6 + 6\sqrt{3})$$

$$A = (36 + 36\sqrt{3}) \text{ units}^2$$

21. $A = \frac{1}{2}h(b_1 + b_2)$

$$96 = \frac{1}{2} \cdot 8(b + 9)$$

$$96 = 4(b + 9)$$

$$96 = 4b + 36$$

$$60 = 4b$$

$$b = 15 \text{ cm}$$

25. Using Heron's Formula, the semiperimeter is
$\frac{1}{2}(3s)$ or $\frac{3s}{2}$. Then

$$A = \sqrt{\frac{3s}{2}\left(\frac{3s}{2} - s\right)\left(\frac{3s}{2} - s\right)\left(\frac{3s}{2} - s\right)}$$

$$A = \sqrt{\frac{3s}{2}\left(\frac{s}{2}\right)\left(\frac{s}{2}\right)\left(\frac{s}{2}\right)}$$

$$A = \sqrt{\frac{3s^4}{16}} = \frac{\sqrt{3} \cdot \sqrt{s^4}}{\sqrt{16}}$$

$$A = \frac{s^2\sqrt{3}}{4}$$

29. In $\triangle ABC$, $AC = 13$ (by using the Pythagorean
Theorem). Now $EC = 12$.

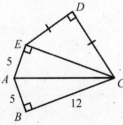

Using the 45°-45°-90° relationship,

$$DE \cdot \sqrt{2} = 12$$

$$DE = \frac{12}{\sqrt{2}} \cdot \frac{\sqrt{2}}{\sqrt{2}} = \frac{12\sqrt{2}}{2} = 6\sqrt{2}$$

$$\therefore DC = 6\sqrt{2} \text{ (since } \overline{DC} \cong \overline{DE}\text{)}$$

$$A_{ABCDE} = A_{ABC} + A_{AEC} + A_{EDC}$$

$$A_{ABCDE} = \frac{1}{2} \cdot 5 \cdot 12 + \frac{1}{2} \cdot 5 \cdot 12 + \frac{1}{2} \cdot (6\sqrt{2})(6\sqrt{2})$$

$$A_{ABCDE} = 30 + 30 + \frac{1}{2}(36)(2)$$

$$A_{ABCDE} = 30 + 30 + 36 = 96 \text{ units}^2$$

33. a. $P = 2b + 2h$
$P = 2(245) + 2(140)$
$P = 490 + 280$
$P = 770 \text{ ft}$

b. Cost = ($0.59)(770)
Cost = $454.30

37. The largest area occurs when the rectangle is a
square with sides of length 10 inches.

41.

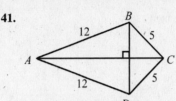

Using Pythagorean Triple (5, 12, 13), $AC = 13$.

$$A_{\triangle ABC} = \sqrt{s(s - a)(s - b)(s - c)} \text{ where}$$

$a = 5$, $b = 12$, $c = 13$ and

$$s = \frac{1}{2}(5 + 12 + 13)$$

$$s = \frac{1}{2}(30) = 15 \ \therefore$$

$$A_{\triangle ABC} = \sqrt{15(15 - 5)(15 - 12)(15 - 13)}$$

$$A_{\triangle ABC} = \sqrt{15(10)(3)(2)}$$

$$A_{\triangle ABC} = \sqrt{(3 \cdot 5) \cdot (2 \cdot 5) \cdot 3 \cdot 2}$$

$$A_{\triangle ABC} = \sqrt{2^2 \cdot 3^2 \cdot 5^2}$$

$$A_{\triangle ABC} = 2 \cdot 3 \cdot 5$$

$$A_{\triangle ABC} = 30 \text{ in}^2$$

Since $A_{\triangle ABC} \cong A_{\triangle ADC}$, then

$$A_{ABCD} = A_{\triangle ABC} + A_{\triangle ADC}$$

$$= 30 + 30 = 60 \text{ in}^2$$

45. The area of a trapezoid $= \frac{1}{2}h(b_1 + b_2)$

$$= h \cdot \frac{1}{2}(b_1 + b_2)$$

$$= h \cdot m$$

since the median, m, of a trapezoid $= \frac{1}{2}(b_1 + b_2)$.

49. $A = \frac{1}{2}d^2 = \frac{1}{2}\left(\sqrt{10}\right)^2 = \frac{1}{2}(10) = 5 \text{ in}^2$

SECTION 8.3: Regular Polygons and Area

1. a. A = s^2 = $(3.5)^2$ = 12.25 cm^2

b. If a = 4.7, then s = 9.4. A = $(9.4)^2$ = 88.36 in^2

5. $c = \dfrac{360}{n}$

$30 = \dfrac{360}{n}$

$n = 12; \ P = 12(5.7) = 68.4$ in.

9. Regular hexagon

13. $A = \dfrac{1}{2}aP$

$A = \dfrac{1}{2}(3.2)(25.6)$

$A = 40.96$ cm^2

17. $A = \dfrac{1}{2}aP = \dfrac{1}{2}(5.2)(37.5) = 97.5$ cm^2

21.

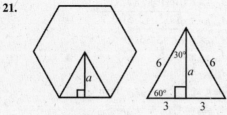

The length of the apothem is $a = 3\sqrt{3}$ (using the 30°-60°-90° relationship).

$A = \dfrac{1}{2}aP$ becomes

$A = \dfrac{1}{2} \cdot 3\sqrt{3} \cdot 36$

$A = 54\sqrt{3}$ cm^2

25. $\dfrac{a}{s} = \dfrac{6}{5} \ \rightarrow 6s = 5a \rightarrow s = \dfrac{5}{6}a$

$P = 8\left(\dfrac{5}{6}a\right) = \dfrac{20}{3}a = \dfrac{20}{3}(15) = 100$

$A = \dfrac{1}{2}P$

$A = \dfrac{1}{2}(15)(100)$

$A = 750$ cm^2

29. $P = 12(2) = 24$

$A = \dfrac{1}{2}aP$

$A = \dfrac{1}{2}(2 + \sqrt{3})24$

$A = 12(2 + \sqrt{3})$

$A = (24 + 12\sqrt{3})$ in^2

33. In octagon RSTUVWXY draw $\overline{SP} \perp \overline{RU}$

$\triangle$ RPS is an isosceles right $\triangle$.

$\therefore$ SP $= \dfrac{4}{\sqrt{2}} \cdot \dfrac{\sqrt{2}}{\sqrt{2}} = \dfrac{4\sqrt{2}}{2} = 2\sqrt{2}$

a, the apothem, $= 2\sqrt{2} + 2$ and RU $= 4 + 4\sqrt{2}$

$A_{RYXWVU} = A_{OCTAGON} - A_{RSTU}$

$= \dfrac{1}{2}aP - \dfrac{1}{2}(b_1 + b_2)$

$= \dfrac{1}{2}(2\sqrt{2} + 2)(32) - \dfrac{1}{2}(2\sqrt{2})(4 + 4\sqrt{2} + 4)$

$= (\sqrt{2} + 1)(32) - \sqrt{2}(8 + 4\sqrt{2})$

$A_{RYXWVU} = 32\sqrt{2} + 32 - 8\sqrt{2} - 8$

$A_{RYXWVU} = 24\sqrt{2} + 24$ units^2

37. In octagon RSTUVWXY draw $\overline{SP} \perp \overline{RU}$

Δ RPS is an isosceles right Δ.

$\therefore$ SP $= \dfrac{4}{\sqrt{2}} \cdot \dfrac{\sqrt{2}}{\sqrt{2}} = \dfrac{4\sqrt{2}}{2} = 2\sqrt{2}$

a, the apothem, $= 2\sqrt{2} + 2$ and RU $= 4 + 4\sqrt{2}$

$A_{RYXWVU} = A_{OCTAGON} - A_{RSTU}$

$= \dfrac{1}{2}aP - \dfrac{1}{2}(b_1 + b_2)$

$= \dfrac{1}{2}\left(2\sqrt{2} + 2\right)(32) - \dfrac{1}{2}\left(2\sqrt{2}\right)\left(4 + 4\sqrt{2} + 4\right)$

$= \left(\sqrt{2} + 1\right)(32) - \sqrt{2}\left(8 + 4\sqrt{2}\right)$

$A_{RYXWVU} = 32\sqrt{2} + 32 - 8\sqrt{2} - 8$

$A_{RYXWVU} = 24\sqrt{2} + 24$ units^2

SECTION 8.4: Circumference and Area of a Circle

1. $\quad C = 2\pi r \qquad A = \pi r^2$

$\quad C = 2 \cdot \pi \cdot 8 \qquad A = \pi \cdot 8^2$

$\quad C = 16\pi$ cm $\quad A = 64\pi$ cm^2

5. a. $\quad C = 2\pi r$

$\quad 44\pi = 2\pi r$

$\quad \dfrac{44\pi}{2\pi} = r$

$\quad r = 22$ in.

$\quad \therefore d = 44$ in.

b. $\quad C = 2\pi r$

$\quad 60\pi = 2\pi r$

$\quad \dfrac{60\pi}{2\pi} = r$

$\quad r = 30$ ft

$\quad \therefore d = 60$ ft

9. $\quad \ell = \dfrac{m}{360} \cdot C$

$\quad \ell = \dfrac{60}{360} \cdot 2 \cdot \pi \cdot 8$

$\quad \ell = \dfrac{1}{6} \cdot 16\pi$

$\quad \ell = \dfrac{8}{3}\pi$ in.

13. $\quad A = \pi r^2$

$\quad 143 = \pi r^2$

$\quad \dfrac{143}{\pi} = r^2$

$\quad r \approx 6.7$ cm

17. The maximum area of a rectangle occurs when it is a square. $\therefore\ A = 16$ sq in.

21. $\quad A = A_{CIRCLE} - A_{SQUARE}$

$\quad A = \pi\left(4\sqrt{2}\right)^2 - 8^2$

$\quad A = (32\pi - 64)$ in^2

25. $\quad A = \pi r^2$

$\quad 154 = \pi r^2$

$\quad r^2 = \dfrac{154}{\pi}$

$\quad r \approx 7$ cm

29. Given: Concentric circles with radii of lengths R and r, where $R > r$.

Explain: $A_{RING} = \pi(R + r)(R - r)$

$\quad A = A_{LARGER\ CIRCLE} - A_{SMALLER\ CIRCLE}$

$\quad A = \pi R^2 - \pi r^2$

$\quad A = \pi\left(R^2 - r^2\right).$

But $R^2 - r^2$ is a difference of 2 squares, so that $A = \pi(R + r)(R - r)$.

33. a. $\quad A = \pi r^2$

$\quad A = \pi\left(8^2\right)$

$\quad A \approx 201.06$ ft^2

b. $\quad \dfrac{201.06}{70} \approx 2.87$ pints

Thus, 3 pints need to be purchased.

c. $\quad 3 \times \$2.95 = \$8.85$

37. $\quad P_{POLYGON} \approx C_{CIRCLE}$

$\quad P \approx 2\pi r$

$\quad P \approx 2 \cdot \pi \cdot 7$

$\quad P \approx 43.98$ cm

41. $\quad C = 2\pi r$

$\quad C = 2\pi(4375)$

$\quad C \approx 27{,}488.94$ miles

45.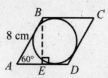

$BE = 4\sqrt{3}$ cm (using the
30°-60°-90° relationship).
$\overline{BE}$ is the same as the diameter of the circle, so
then the radius of the circle $= 2\sqrt{3}$ cm .

$$A_{\text{CIRCLE}} = \pi r^2 = \pi\left(2\sqrt{3}\right)^2 = \pi \cdot 4 \cdot 3 = 12\pi \text{ cm}^2$$

SECTION 8.5: More Area Relationships in the Circle

1. $P_{\text{SECTOR}} = 10 + 10 + 14 = 34$ in.

5. $A_\triangle = \dfrac{1}{2}rP$

$A_\triangle = \dfrac{1}{2}r(3s)$

$A_\triangle = \dfrac{3}{2}rs$

9. $A_\triangle = \dfrac{1}{2}rP$

$A_\triangle = \dfrac{1}{2}(2)(6+8+10)$

$A_\triangle = 24$ sq in.

13. $P = 2r + \ell$

$P = 2(8) + \dfrac{8}{3}\pi$

$P = \left(16 + \dfrac{8}{3}\pi\right)$ in.

$A = \dfrac{m}{360} \cdot \pi r^2$

$A = \dfrac{60}{360} \cdot \pi\left(8^2\right)$

$A = \dfrac{1}{6} \cdot 64\pi$

$A = \dfrac{32}{3}\pi$ in^2

17. $P = AB + \ell\overset{\frown}{AB}$

$P = 12 + \dfrac{60}{360} \cdot 2\pi(12)$

$P = 12 + \dfrac{1}{6}(24\pi)$

$P = (12 + 4\pi)$ in.

$A = A_{\text{SECTOR}} - A_\triangle$

$A = \dfrac{1}{6}(\pi)(12^2) - \dfrac{12^2}{4}\sqrt{3}$

$A = \dfrac{1}{6}(144\pi) - \dfrac{144}{4}\sqrt{3}$

$A = \left(24\pi - 36\sqrt{3}\right)$ in^2

21. $A = \dfrac{m}{360} \cdot \pi r^2$

$\dfrac{9}{4}\pi = \dfrac{40}{360} \cdot \pi r^2$

$\dfrac{9}{4}\pi = \dfrac{1}{9}\pi r^2$

Multiply by 9

$\dfrac{81}{4}\pi = \pi r^2$

$r^2 = \dfrac{81}{4}$

$r = \dfrac{9}{2}$ cm

25. $\ell = \dfrac{m}{360} \cdot 2\pi r$

$6\pi = \dfrac{m}{360} \cdot 2\pi(12)$

$6\pi = \dfrac{m}{360} \cdot 24\pi$

Dividing by 24π, we get

$\dfrac{1}{4} = \dfrac{m}{360}$

Multiplying by 360, we get $m = 90°$.

29. Draw in two radii to consecutive vertices of the square. Then

$r^2 + r^2 = s^2$

$2r^2 = s^2$

$r^2 = \dfrac{s^2}{2}$

$r = \dfrac{s}{\sqrt{2}} = \dfrac{s\sqrt{2}}{2}$

$A = A_{\text{CIRCLE}} - A_{\text{SQUARE}}$

$A = \pi r^2 - s^2$

$A = \pi\left(\dfrac{s\sqrt{2}}{2}\right)^2 - s^2$

$A = \pi \cdot \dfrac{s^2 \cdot 2}{4} - s^2$

$A = \left(\dfrac{\pi}{2}\right)s^2 - s^2$

33. $A_\triangle = \sqrt{s(s-a)(s-b)(s-c)}$

Also, $A_\triangle = \dfrac{1}{2}rP$ or $A_\triangle = \dfrac{1}{2}r(a+b+c)$.

So $\dfrac{1}{2}r(a+b+c) = \sqrt{s(s-a)(s-b)(s-c)}$.

Solve for r.

$r = \dfrac{2\sqrt{s(s-a)(s-b)(s-c)}}{a+b+c}$

37. $A = \dfrac{120}{360} \cdot \pi \left(18^2 - 4^2\right)$

$A = \dfrac{1}{3} \cdot \pi \cdot (324 - 16)$

$A = \dfrac{\pi}{3}(308)$

$A = \dfrac{308\pi}{3} \approx 322.54$ sq in.

CHAPTER REVIEW

1.

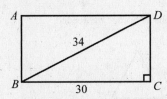

Using the Pythagorean Theorem,

$(34)^2 = (30)^2 + (DC)^2$

$1156 = 900 + (DC)^2$

$256 = (DC)^2$

$DC = 16$

$A = 30(16) = 480$ units^2

2. a. $A = 10(4) = 40$ units^2

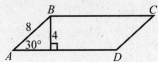

b. $A = 10\left(4\sqrt{3}\right) = 40\sqrt{3}$ units^2

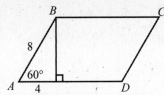

c. $A = 10\left(4\sqrt{2}\right) = 40\sqrt{2}$ units^2

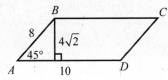

3. Using the 45°-45°-90° relationship,

$AB = DB = 5\sqrt{2}$.

$\therefore A_{ABCD} = bh$

$A_{ABCD} = \left(5\sqrt{2}\right)\left(5\sqrt{2}\right)$

$A_{ABCD} = 50$ units^2

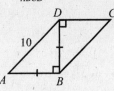

4. $A = \sqrt{s(s-a)(s-b)(s-c)}$ where

$s = \dfrac{1}{2}(17 + 25 + 26)$

$s = \dfrac{1}{2}(68)$

$s = 34$

$A = \sqrt{34(34 - 26)(34 - 25)(34 - 17)}$

$A = \sqrt{34(8)(9)(17)}$

$A = \sqrt{2^4 \cdot 3^2 \cdot 17^2}$

$A = 2^2 \cdot 3 \cdot 17$

$A = 204$ units^2

5. $A = \sqrt{s(s-a)(s-b)(s-c)}$ where

$s = \dfrac{1}{2}(26 + 28 + 30)$

$s = \dfrac{1}{2}(84)$

$s = 42$

$A = \sqrt{42(42 - 26)(42 - 28)(42 - 30)}$

$A = \sqrt{42(16)(14)(12)}$

$A = \sqrt{2^8 \cdot 3^2 \cdot 7^2}$

$A = 2^4 \cdot 3 \cdot 7$

$A = 336$ units^2

6. Using the (3, 4, 5) Pythagorean Triple,
$BH = 4$

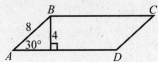

$A = \dfrac{1}{2} h \left(b_1 + b_2\right)$

$A = \dfrac{1}{2} \cdot 4(6 + 12)$

$A = 36$ units^2

7.

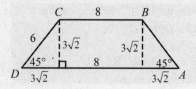

a. $A = \frac{1}{2}h(b_1 + b_2)$

$A = \frac{1}{2} \cdot 3\sqrt{2}\left(8 + (8 + 6\sqrt{2})\right)$

$A = \frac{1}{2} \cdot 3\sqrt{2}\left(16 + 6\sqrt{2}\right)$

$A = 3\sqrt{2}\left(8 + 3\sqrt{2}\right)$

$A = \left(24\sqrt{2} + 18\right)$ units^2

b. $A = \frac{1}{2} \cdot 3\left(8 + (8 + 6\sqrt{3})\right)$

$A = \frac{3}{2}\left(16 + 6\sqrt{3}\right)$

$A = \left(24 + 9\sqrt{3}\right)$ units^2

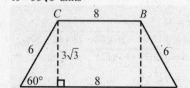

c. $A = \frac{1}{2} \cdot 3\sqrt{3}\left(8 + 14\right)$

$A = \frac{1}{2}(3\sqrt{3})(22)$

$A = 33\sqrt{3}$ units^2

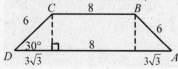

8. $A = \frac{1}{2}d_1 \cdot d_2$

$A = \frac{1}{2}(18)(24)$

$A = 216$ in^2

Using the (9, 12, 15) Pythagorean Triple,
$P = 4(15) = 60$ in.

9. a. $A = (140)(160) - \left[(80)(35) + (30)(20)\right]$

$A = 22,400 - \left[2800 + 600\right]$

$A = 22,400 - 3400$

$A = 19,000$ ft^2

b. $\frac{19,000}{5000} = 3.8$ bags

Tom needs to buy 4 bags.

c. Cost $= 4 \cdot \$18 = \$72$

10. a. $A = (9 \cdot 8) + (12 \cdot 8)$

$A = 72 + 96$

$A = 168$ ft^2

$\frac{168}{60} = 2.8$ double rolls

3 double rolls are needed.

b. $P = 2a + 2b$

$P = 2(9) + 2(12)$

$P = 18 + 24$

$P = 42$ ft or 14 yd

$\frac{14}{5} = 2.8$ rolls

3 rolls of border are needed.

11.

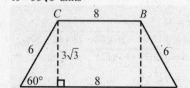

If $\triangle FBC$ is equilateral, then so is $\triangle FAD$.

$\therefore \quad AD = 17$

$(AD)^2 = (AE)^2 + (ED)^2$

$(17)^2 = (AE)^2 + (16)^2$

$28 = (AE)^2 + 256$

$(AE)^2 = 33$

$AE = \sqrt{33}$

a. $A_{EAFD} = A_{FAD} + A_{AED}$

$A_{EAFD} = \frac{17^2}{4}\sqrt{3} + \frac{1}{2}(16)\sqrt{33}$

$A_{EAFD} = \left(\frac{289}{4}\sqrt{3} + 8\sqrt{33}\right)$ units^2

b. $P_{EAFD} = AF + FD + DE + AE$

$P_{EAFD} = 17 + 17 + 16 + \sqrt{33}$

$P_{EAFD} = \left(50 + \sqrt{33}\right)$ units

12.

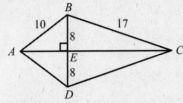

$AE = 6$
(using the 6-8-10 Pythagorean Triple)
$EC = 15$
(using the 8-15-17 Pythagorean Triple)
$A = \frac{1}{2} d_1 \cdot d_2$
$A = \frac{1}{2}(BD)(AC)$
$A = \frac{1}{2}(16)(6+15)$
$A = 8(21)$
$A = 168 \text{ units}^2$

13.

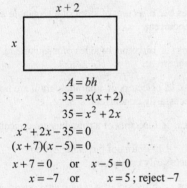

$A = bh$
$35 = x(x+2)$
$35 = x^2 + 2x$
$x^2 + 2x - 35 = 0$
$(x+7)(x-5) = 0$
$x+7 = 0 \quad \text{or} \quad x-5 = 0$
$\quad x = -7 \quad \text{or} \quad x = 5 \text{ ; reject } -7$
The rectangle is 5 cm by 7 cm.

14. $P = a + b + c$

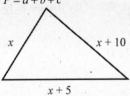

a. $60 = x + (x+10) + (x+5)$
$60 = 3x + 15$
$3x = 45$
$x = 15$
$x + 10 = 25$
$x + 5 = 20$
The three sides have lengths 15 cm, 25 cm, and 20 cm.

b. (15, 20, 25) is a Pythagorean Triple in which 25 is the length of the hypotenuse.
$A = \frac{1}{2} bh$
$A = \frac{1}{2}(20)(15)$
$A = 150 \text{ cm}^2$

15.

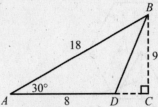

Using the 30°-60°-90° relationship, $BC = 9$.
$A = \frac{1}{2} bh$
$A = \frac{1}{2}(8)(9)$
$A = 36 \text{ units}^2$

16. $A = \frac{s^2}{4}\sqrt{3}$
$A = \frac{12^2}{4}\sqrt{3}$
$A = \frac{144}{4}\sqrt{3}$
$A = 36\sqrt{3} \text{ cm}^2$

17.

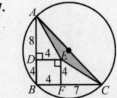

$A = A_{ABC} - (A_{ADE} + A_{BDEF} + A_{EFC})$
$A = \frac{1}{2}(11)(12) - \left[\frac{1}{2}(8)(4) + 4^2 + \frac{1}{2}(4)(7) \right]$
$A = 66 - [16 + 16 + 14]$
$A = 66 - 46$
$A = 20 \text{ units}^2$

18. a. $c = \frac{360}{5} = 72°$

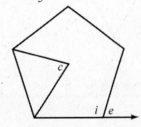

b. $i = \frac{(5-2)180}{2} = \frac{3(180)}{5} = \frac{540}{5} = 180°$

c. $e = 180° - 108° = 72°$

19.

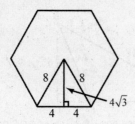

$$A = \frac{1}{2}aP$$
$$A = \frac{1}{2}(4\sqrt{3})(6 \cdot 8)$$
$$A = 96\sqrt{3} \text{ ft}^2$$

20. $P = 3(12\sqrt{3}) = 36\sqrt{3}$

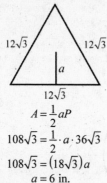

$$A = \frac{1}{2}aP$$
$$108\sqrt{3} = \frac{1}{2} \cdot a \cdot 36\sqrt{3}$$
$$108\sqrt{3} = (18\sqrt{3})a$$
$$a = 6 \text{ in.}$$

21. $x \cdot \sqrt{3} = 9$
$$x = \frac{9}{\sqrt{3}}$$
$$x = \frac{9}{\sqrt{3}} \cdot \frac{\sqrt{3}}{\sqrt{3}} = \frac{9\sqrt{3}}{3} = 3\sqrt{3}$$

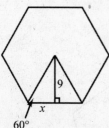

Each side has length $6\sqrt{3}$ in.
$$P = n(s)$$
$$P = 6(6\sqrt{3})$$
$$P = 36\sqrt{3} \text{ in.}$$
$$A = \frac{1}{2}aP$$
$$A = \frac{1}{2}(9)(36\sqrt{3})$$
$$A = 162\sqrt{3} \text{ in}^2$$

22. a. $c = \dfrac{360}{n}$
$$45 = \frac{360}{n}$$
$$45n = 360$$
$$n = 8$$

b. $P = n(s)$
$$P = 8(5)$$
$$P = 40$$
$$A = \frac{1}{2}aP$$
$$A \approx \frac{1}{2}(6)(40)$$
$$A \approx 120 \text{ cm}^2$$

23. a. No. $\perp$ bisectors of sides of a parallelogram are not necessarily concurrent.

b. No. $\perp$ bisectors of sides of a rhombus are not concurrent.

c. Yes. $\perp$ bisectors of sides of a rectangle are concurrent.

d. Yes. $\perp$ bisectors of sides of a square are concurrent.

24. a. No. $\angle$ bisectors of a parallelogram are not necessarily concurrent.

b. Yes. $\angle$ bisectors of a rhombus are concurrent.

c. No. $\angle$ bisectors of a rectangle are not necessarily concurrent.

d. Yes. $\angle$ bisectors of a square are concurrent.

25. If the radius of the inscribed circle is 7 in., then the length of each side of the triangle is $14\sqrt{3}$ in.
$$A_\triangle = \frac{1}{2}rP$$
$$A_\triangle = \frac{1}{2}(7)(3 \cdot 14\sqrt{3})$$
$$A_\triangle = 147\sqrt{3}$$
$$A_\triangle \approx 254.61 \text{ sq in.}$$

26. a. $A = (20)(30) - (12)(24)$
$$A = 600 - 288$$
$$A = 312 \text{ ft}^2$$

b. $\dfrac{312}{9} = 34\dfrac{2}{3} \text{ yd}^2$

35 yd^2 should be purchased.

c. $35 \cdot \$9.97 = \$348.95$

27. $A = A_{SQUARE} - 4 \cdot A_{SECTOR}$

$A = 8^2 - 4 \cdot \left[\frac{90}{360} \left(\pi \cdot 4^2 \right) \right]$

$A = 64 - 4 \left[\frac{1}{4} (16\pi) \right]$

$A = (64 - 16\pi) \text{ units}^2$

28. $A = A_{SEMICIRCLE} - A_{TRIANGLE}$

$A = \frac{1}{2} \left(\pi \cdot 7^2 \right) - \frac{1}{2} (7) \left(7\sqrt{3} \right)$

$A = \left(\frac{49}{2} \pi - \frac{49}{2} \sqrt{3} \right) \text{ units}^2$

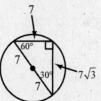

29. $A_{SEGMENT} = A_{SECTOR} - A_{TRIANGLE}$

$A_{SEGMENT} = \frac{60}{360} \cdot \pi \cdot 4^2 - \frac{4^2}{4} \sqrt{3}$

$A_{SEGMENT} = \left(\frac{8}{3} \pi - 4\sqrt{3} \right) \text{ units}^2$

30.

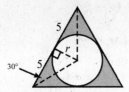

$A = A_{RECT} - 2 \cdot A_{CIRCLE}$

$A = (12)(24) - 2 \cdot \left(\pi \cdot 6^2 \right)$

$A = (288 - 72\pi) \text{ units}^2$

31.

The radius of the circle is r.

$r\sqrt{3} = 5$

$r = \frac{5}{\sqrt{3}} = \frac{5\sqrt{3}}{3}$

$A = A_{EQ.\triangle} - A_{CIRCLE}$

$A = \frac{10^2}{4} \sqrt{3} - \pi \left(\frac{5\sqrt{3}}{3} \right)^2$

$A = 25\sqrt{3} - \pi \cdot \frac{75}{9}$

$A = \left(25\sqrt{3} - \frac{25}{3} \pi \right) \text{ units}^2$

32.

$\ell = \frac{m}{360} (2\pi r)$

$\ell = \frac{40}{360} \left(2 \cdot \pi \cdot 3\sqrt{5} \right)$

$\ell = \frac{1}{9} \left(6\pi\sqrt{5} \right)$

$\ell = \frac{2\pi\sqrt{5}}{3} \text{ cm}$

$A = \frac{m}{360} \left(\pi r^2 \right)$

$A = \frac{40}{360} \left(\pi \cdot \left(3\sqrt{5} \right)^2 \right)$

$A = \frac{1}{9} (\pi \cdot 45)$

$A = 5\pi \text{ cm}^2$

33. **a.** $C = \pi d$

$66 = \frac{22}{7} \cdot d$

$\frac{7}{22} \cdot 66 = \frac{7}{22} \cdot \frac{22}{7} d$

$d = 21 \text{ ft}$

b. $d = 21$, $r = \frac{21}{2}$

$A = \pi r^2$

$A = \frac{22}{7} \left(\frac{21}{2} \right)^2$

$A = \frac{693}{2} \approx 346\frac{1}{2} \text{ ft}^2$

34. **a.** $A_{SECTOR} = \frac{m}{360} \left(\pi r^2 \right)$

$A = \frac{80}{360} (27\pi)$

$A = \frac{2}{9} (27\pi)$

$A = 6\pi \text{ ft}^2$

b. Since $\pi r^2 = 27\pi$,

$$r^2 = 27$$
$$r = \sqrt{27}$$
$$r = 3\sqrt{3}$$

$$\ell = \frac{m}{360}(2\pi r)$$
$$\ell = \frac{80}{360}(2\pi \cdot 3\sqrt{3})$$
$$\ell = \frac{2}{9}(6\pi\sqrt{3})$$
$$\ell = \frac{4\pi}{3}\sqrt{3}$$
$$P = 2r + \ell$$
$$P = 2(3\sqrt{3}) + \frac{4\pi}{3}\sqrt{3}$$
$$P = \left(6\sqrt{3} + \frac{4\pi}{3}\sqrt{3}\right) \text{ ft}$$

35.

$$A_{\text{SHADED}} = A_{\text{SEMICIRCLE}} - A_\triangle$$
$$A_{\text{SHADED}} = \frac{1}{2} \cdot (\pi \cdot 6^2) - \frac{1}{2} \cdot 6 \cdot 12$$
$$A_{\text{SHADED}} = 18\pi - 36$$

The area sought is one-half the shaded area.

$$A = \frac{1}{2}(18\pi - 36)$$
$$A = (9\pi - 18) \text{ in}^2$$

36. Given: Concentric circles with radii of lengths R and r with $R > r$; O is the center of the circles.

Prove: $A_{\text{RING}} = \pi(BC)^2$

Proof: By an earlier theorem,

$$A_{\text{RING}} = \pi R^2 - \pi r^2$$
$$= \pi(OC)^2 - \pi(OB)^2$$
$$= \pi\left[(OC)^2 - (OB)^2\right]$$

In rt. $\triangle OBC$,

$$(OB)^2 + (BC)^2 = (OC)^2$$
$$\therefore (OC)^2 - (OB)^2 = (BC)^2$$

In turn, $A_{\text{RING}} = \pi(BC)^2$

37. The area of a circle circumscribed about a square is twice the area of the circle inscribed within the square.

Proof: Let r represent the length of radius of the inscribed circle.
Using the 45°-45°-90° relationship, the radius of the larger circle is $r\sqrt{2}$.
Now,

$$A_{\text{INSCRIBED CIRCLE}} = \pi r^2$$
$$A_{\text{CIRCUMSCRIBED CIRCLE}} = \pi\left(r\sqrt{2}\right)^2$$
$$= \pi \cdot (r^2 \cdot 2)$$
$$= 2(\pi r^2)$$
$$= 2(A_{\text{INSCRIBED CIRCLE}})$$

38. If semicircles are constructed on each of the sides of a right triangle, then the area of the semicircle on the hypotenuse is equal to the sum of the areas of the semicircles on the two legs.

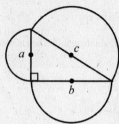

Proof: The radii of the semicircles are $\frac{1}{2}a$, $\frac{1}{2}b$,

and $\frac{1}{2}c$. The area of the semicircle on the

hypotenuse is

$A = \frac{1}{2}\left[\pi\left(\frac{1}{2}c\right)^2\right]$

$= \frac{1}{2}\cdot\pi\left(\frac{1}{4}c^2\right)$

$= \frac{1}{2}\pi\left[\frac{1}{4}(a^2 + b^2)\right]$ $(c^2 = a^2 + b^2$ by the

Pythagorean Theorem.)

$= \frac{1}{2}\pi\left[\frac{1}{4}a^2 + \frac{1}{4}b^2\right]$

$= \frac{1}{2}\pi\left(\frac{1}{4}a^2\right) + \frac{1}{2}\pi\left(\frac{1}{4}b^2\right)$

$= \frac{1}{2}\pi\left(\frac{1}{2}a\right)^2 + \frac{1}{2}\pi\left(\frac{1}{2}b\right)^2$

$= \frac{1}{2}\left[\pi\left(\frac{1}{2}a\right)^2\right] + \frac{1}{2}\left[\pi\left(\frac{1}{2}b\right)^2\right]$

which is the sum of areas of the semicircles on the two legs.

39. a. $A = (18)(15) - \left[\frac{1}{2}(3.14)\cdot 3^2 + \frac{1}{4}(3.14)\cdot 3^2\right]$

$A = 270 - 21.2$

$A = 248.8 \text{ ft}^2$

The approx. number of yd^2 of carpeting

needed is $\frac{248.8}{9} \approx 28$.

b. From (a), the number of ft^2 to be tiled is

21.2 ft^2.

40. a. $A = \frac{1}{2}\left[\pi R^2 - \pi r^2\right]$

$A = \frac{1}{2}\left[(3.14)(30)^2 - (3.14)(18)^2\right]$

$A = \frac{1}{2}[1808.64]$

$A = 904.32 \text{ ft}^2 \approx 905 \text{ ft}^2$

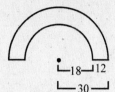

b. Cost = (905)($0.18)

Cost = $162.90

c. Length $= \frac{1}{2}(2\pi R) + \frac{1}{2}(2\pi r)$

$= \frac{1}{2}[2(3.14)(31)] + \frac{1}{2}[2(3.14)(17)]$

$= 97.34 + 52.38$

$= 150.72$

CHAPTER TEST

1. a. square inches

b. equal

2. a. $A = s^2$

b. $C = 2\pi r$

3. a. True

b. False

4. 23 cm^2

5.

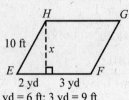

$$2 \text{ yd} = 6 \text{ ft}; 3 \text{ yd} = 9 \text{ ft}$$
$$10^2 = 6^2 + x^2$$
$$x^2 = 10^2 - 6^2$$
$$x = \sqrt{100 - 36}$$
$$x = \sqrt{64} = 8$$
$$A_{EFGH} = bh = (6+9)(8) = 120 \text{ ft}^2$$

6. $A_{MNPQ} = \frac{1}{2}d_1 d_2 = \frac{1}{2}(8)(6) = 24 \text{ ft}^2$

7. $s = \frac{1}{2}(4 + 13 + 15) = 16$

$$A = \sqrt{s(s-a)(s-b)(s-c)}$$
$$A = \sqrt{16(16-4)(16-13)(16-15)}$$
$$A = \sqrt{16(12)(3)(1)}$$
$$A = \sqrt{4^2 \cdot 3^2 \cdot 2^2}$$
$$A = 4 \cdot 3 \cdot 2$$
$$A = 24 \text{ cm}^2$$

8.

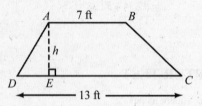

$$A = \frac{1}{2}h(b_1 + b_2)$$
$$60 = \frac{1}{2}h(7 + 13)$$
$$120 = 20h$$
$$h = 6$$
$$AE = 6 \text{ ft}$$

9. a. $P = ns = 5(5.8) = 29 \text{ in.}$

b. $A = \frac{1}{2}aP = \frac{1}{2}(4.0)(29) = 58 \text{ in}^2$

10. a. $C = 2\pi r = 2\pi(5) = 10\pi \text{ in.}$

b. $A = \pi r^2 = \pi(5)^2 = 25\pi \text{ in}^2$

11. $\ell \widehat{AC} = \frac{m\widehat{AC}}{360} \cdot 2\pi r$

$$\ell \widehat{AC} = \frac{45}{360} \cdot 2\left(\frac{22}{7}\right)(7)$$
$$\ell \widehat{AC} = \frac{1}{8} \cdot 44$$
$$\ell \widehat{AC} = \frac{11}{2} = 5\frac{1}{2} \text{ in.}$$

12. $d = 2r$
$$20 = 2r$$
$$r = 10 \text{ cm}$$
$$A = \pi r^2$$
$$A = (3.14)(10)^2 = 314 \text{ cm}^2$$

13. By the 45°-45°-90° relationship, the radius of the circle is 4.

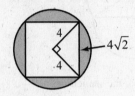

$$A_{SHADED} = A_{CIRCLE} - A_{SQUARE}$$
$$A_{SHADED} = (\pi \cdot r^2) - s^2$$
$$A_{SHADED} = (\pi \cdot 4^2) - (4\sqrt{2})^2$$
$$A_{SHADED} = (16\pi - 32) \text{ in}^2$$

14. $A = \frac{m}{360}\pi r^2$

$$A = \frac{135}{360} \cdot \pi \cdot 12^2$$
$$A = \frac{3}{8} \cdot \pi \cdot 144$$
$$A = 54\pi \text{ cm}^2$$

15.

$$A_{SHADED} = A_{SECTOR} - A_{TRIANGLE}$$
$$A_{SHADED} = \frac{m}{360}\pi r^2 - \frac{1}{2}bh$$
$$A_{SHADED} = \frac{90}{360} \cdot \pi \cdot 12^2 - \frac{1}{2}(12)(12)$$
$$A_{SHADED} = \frac{1}{4} \cdot \pi \cdot 144 - 72$$
$$A_{SHADED} = (36\pi - 72) \text{ in}^2$$

16. $P = a + b + c$
$$P = 5 + 12 + 13$$
$$P = 30$$
$$A = \frac{1}{2}rP$$
$$30 = \frac{1}{2}r(30)$$
$$1 = \frac{1}{2}r$$
$$r = 2 \text{ in.}$$

Chapter 9: Surfaces and Solids

SECTION 9.1: Prisms, Area, and Volume

1. a. Yes

    b. Oblique

    c. Hexagon

    d. Oblique hexagonal prism

    e. Parallelogram

5. a. cm^2

    b. cm^3

9. $V = bh$
$V = 12(10)$
$V = 120 \ cm^3$

13. a. $2n$

    b. n

    c. $2n$

    d. $3n$

    e. n

    f. 2

    g. $n + 2$

17. a. $L = hP$
$L = (6)(3 + 4 + 5)$
$L = 72 \ ft^2$

    b. $T = L + 2B$
$T = 72 + 2\left(\dfrac{1}{2}bh\right)$
$T = 72 + 4 \cdot 3$
$T = 84 \ ft^2$

    c. $V = Bh$
$V = \left(\dfrac{1}{2}bh\right) \cdot 6$
$V = \left(\dfrac{1}{2} \cdot 4 \cdot 3\right)(6)$
$V = 36 \ ft^3$

21.

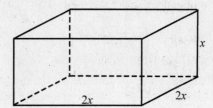

$V = l \cdot w \cdot h$
$108 = 2x \cdot 2x \cdot x$
$108 = 4x^3$
$x^3 = 27$
$x = 3 \rightarrow 2x = 6$

25. 1 foot equals 12 inches.
$L = 5 \cdot 6 + 12 \cdot 6 + 5 \cdot 6 + 12 \cdot 6$
$L = 30 + 72 + 30 + 7$
$L = 204 \ in^2$
$B = 5 \cdot 12 = 60 \ in^2$
The area of both bases is 120 in^2.
Cost $= \$0.01(204) + \$0.02(120)$
Cost $= \$2.04 + \$2.40$
Cost $= \$4.44$

29. a. $T = L + 2B$
$T = hP + 2(e \cdot e)$
$T = e(4e) + 2e^2$
$T = 4e^2 + 2e^2$
$T = 6e^2$

    b. $T = 6e^2$
$T = 6 \cdot 4^2$
$T = 96 \ cm^2$

    c. $V = Bh$
$V = e^2 \cdot e$
$V = e^3$

    d. $V = e^3$
$V = 4^3 = 64 \ cm^3$

35. $V = 15' \cdot 12' \cdot 2'$
$V = 360 \ ft^3$
$1 \ yd^3 = 27 \ ft^3$
$V = 360 \ ft^3 \div \dfrac{27 \ ft^3}{1 \ yd^3}$
$V = 13\dfrac{1}{3} \ yd^3$
Cost $= 13\dfrac{1}{3} \ yd^3 \cdot \dfrac{\$9.60}{1 \ yd^3} = \$128$

39. $V = 2 \cdot 1 \cdot \dfrac{2}{3} = \dfrac{4}{3}$ ft^3

# of gallons $= \dfrac{4}{3} \cdot 7.5 = 10$ gal

43. $V = Bh$

$V = \left(\dfrac{1}{2}aP\right) \cdot 12$

$V = \dfrac{1}{2} \cdot 8.2 \cdot (12 \cdot 5) \cdot 12$

$V = 2952$ cm^3

SECTION 9.2: Pyramids, Area, and Volume

1. a. Right pentagonal prism

b. Oblique pentagonal prism

5. a. Pyramid

b. E

c. $\overline{EA}$, $\overline{EB}$, $\overline{EC}$, $\overline{ED}$

d. $\triangle EAB$, $\triangle EBC$, $\triangle ECD$, $\triangle EAD$

e. No

9. $T = 12 + 16 + 12 + 10 + 16$

$T = 66$ in^2

13. a. $n + 1$

b. n

c. n

d. $2n$

e. n

f. $n + 1$

17. a. Slant height

b. Lateral edge

21. a. $B = \dfrac{1}{2}aP$

$B = \dfrac{1}{2}(6.3)(5 \cdot 9.2)$

$B = 144.9$ cm^2

b. $V = \dfrac{1}{3}Bh$

$V = \dfrac{1}{3}(144.9)(14.6)$

$V = 705.18$ cm^3

25. $V = \dfrac{1}{3}Bh$

$72 = \dfrac{1}{3} \cdot x^2 \cdot x$

$72 = \dfrac{1}{3}x^3$

$x^3 = 216$

$x = 6$

$l^2 = a^2 + h^2$

$l^2 = 3^2 + 6^2$

$l^2 = 45$

$l = \sqrt{45} = 3\sqrt{5}$

$L = \dfrac{1}{2}lP$

$L = \dfrac{1}{2}(3\sqrt{5})(4 \cdot 6)$

$L = 36\sqrt{5}$

$T = L + B$

$T = 36\sqrt{5} + (6 \cdot 6)$

$T = 36\sqrt{5} + 36 \approx 116.5$ in^2

29. $V = \dfrac{1}{3}Bh$

$V = \dfrac{1}{3}\left(\dfrac{1}{2}aP\right)(15)$

$V = \dfrac{1}{3} \cdot \dfrac{1}{2} \cdot (7.5)(12 \cdot 4)(15)$

$V = 900$ ft^3

33. $V = V_{\text{TALLER PYRAMID}} - V_{\text{SHORTER PYRAMID}}$

$V = \dfrac{1}{3}B_1h_1 - \dfrac{1}{3}B_2h_2$

$V = \dfrac{1}{3}(6 \cdot 6)(32) - \dfrac{1}{3}(3 \cdot 3)(16)$

$V = 384 - 48$

$V = 336$ in^3

37

$l^2 + 9 = 34$

$l^2 = 25$

$l = 5$

$T = B + L$

$T = 36 + \dfrac{1}{2} \cdot 24 \cdot 5$

$T = 36 + 60$

$T = 96$ in^2

41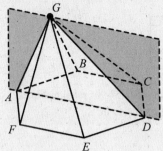

$V_{ABCD} = V_{DEFA}$
$V_{ABCDEF} = V_{ABCD} + V_{DEFA}$
$V_{ABCDEF} = 19.7 + 19.7 = 39.4 \text{ in}^3$

SECTION 9.3: Cylinders and Cones

1. **a.** Yes

   b. Yes

   c. Yes

5. $L = 2\pi rh$
   $L = 2\pi(1.5)(4.25)$
   $L = 12.75\pi \text{ in}^2$

   $B = \pi r^2$
   $B = \pi(1.5)^2$
   $B = 2.25\pi$

   $T = L + 2B$
   $T = 12.75\pi + 2(2.25\pi)$
   $T = 17.25\pi \approx 54.19 \text{ in}^3$

9. $L = 2\pi rh$
   $12\pi = 2\pi r(4+1)$
   $12\pi = 2\pi(r^2 + r)$
   $\dfrac{12\pi}{2\pi} = \dfrac{2\pi(r^2 + r)}{2\pi}$
   $r^2 + r = 6$
   $r^2 + r - 6 = 0$
   $(r+3)(r-2) = 0$
   $r + 3 = 0$ or $r - 2 = 0$
   $r = -3$ or $r = 2$; reject -3

   The radius has a length of 2 inches and the altitude has a length of 3 inches.

13. $r^2 + h^2 = l^2$
    $4^2 + 6^2 = l^2$
    $l^2 = 52$
    $l = \sqrt{52} = 2\sqrt{13} \approx 7.21 \text{ cm}$

17. $r^2 + h^2 = l^2$
    $6^2 + h^2 = (2h)^2$
    $3h^2 = 36$
    $h^2 = 12$
    $h = 2\sqrt{3}$
    $l = 2h = 2(2\sqrt{3}) = 4\sqrt{3} \approx 6.93 \text{ in.}$

21. $l^2 = r^2 + h^2$
    $l^2 = 7^2 + 6^2$
    $l^2 = 85$
    $l = \sqrt{85}$

    a. $L = \dfrac{1}{2} lC$
    $L = \dfrac{1}{2}\sqrt{85} \cdot 2\pi r$
    $L = \dfrac{1}{2}\sqrt{85} \cdot 2 \cdot \pi \cdot 6$
    $L = 6\pi\sqrt{85} \approx 173.78 \text{ in}^2$

    b. $T = L + B$
    $T = 6\pi\sqrt{85} + \pi r^2$
    $T = 6\pi\sqrt{85} + \pi \cdot 6^2$
    $T = 6\pi\sqrt{85} + 36\pi \approx 286.88 \text{ in}^2$

    c. $V = \dfrac{1}{3} Bh$
    $V = \dfrac{1}{3}\pi r^2 \cdot 7$
    $V = \dfrac{1}{3} \cdot \pi \cdot 6^2 \cdot 7$
    $V = 84\pi \approx 263.89 \text{ in}^3$

25. The solid formed is a cone with $r = 20$ and $h = 15$.
    $V = \dfrac{1}{3} Bh$
    $V = \dfrac{1}{3}\pi r^2 h$
    $V = \dfrac{1}{3} \cdot \pi \cdot 20^2 \cdot 15$
    $V = 2000\pi$

29. $d = 10 \therefore r = 5$

$$V = \frac{1}{3}Bh$$

$$V = \frac{1}{3}\pi r^2 h$$

$$100\pi = \frac{1}{3} \cdot \pi \cdot 5^2 \cdot h$$

$$100\pi = \frac{25\pi}{3} \cdot h$$

$$h = 12$$

$$L = \frac{1}{2}lC$$

$$L = \frac{1}{2} \cdot 13 \cdot 2\pi r$$

$$L = \frac{1}{2} \cdot 13 \cdot 2\pi \cdot 5$$

$$L = 65\pi \approx 204.2 \text{ cm}^2$$

33. $V = V_{\text{LARGE CYLINDER}} - V_{\text{SMALL CYLINDER}}$

$$V = \pi R^2 h - \pi r^2 h$$

$$V = \pi h(R^2 - r^2)$$

$$V = \pi h(R - r)(R + r)$$

37. Lateral area of smaller $= 2\pi rh$

Lateral area of larger $= 2\pi(2r)(2h) = 8\pi rh$

$$\frac{\text{Lateral area of larger}}{\text{Lateral area of smaller}} = \frac{8\pi rh}{2\pi rh} = \frac{4}{1} \text{ or } 4{:}1$$

41. $V = V_{\text{LARGER CONE}} - V_{\text{SMALLER CONE}}$

$$V = \frac{1}{3}\pi R^2 h - \frac{1}{3}\pi r^2 h$$

45. $V = \pi r^2 h$

$$V = \pi \cdot (1.5)^2 \cdot 6$$

$$V = 13.5\pi$$

$$V \approx 42.4115 \text{ ft}^3$$

$$\text{\# of gallons} = 42.4115 \cdot 7.5 \approx 318 \text{ gal}$$

SECTION 9.4: Polyhedrons and Spheres

1. Polyhedron *EFGHIJK* is concave.

5. A regular hexahedron has 6 faces (F), 8 vertices (V), and 12 edges (E).

$$V + F = E + 2$$

$$8 + 6 = 12 + 2$$

9. Nine faces

13.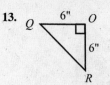

a. Using the 45°-45°-90° relationship,

$$QR = 6\sqrt{2} \approx 8.49 \text{ in.}$$

b.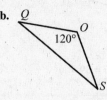

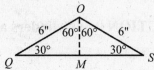

Where M is the midpoint of $\overline{QS}$, $OM = 3$ and $QM = 3\sqrt{3}$ by the 30°-60°-90° relationship.

$$QS = 2(QM) = 2(3\sqrt{3}) = 6\sqrt{3} \ .$$

That is, $QS = 6\sqrt{3} \approx 10.39 \text{ in.}$

17. $S = 6(4.2)^2 = 105.84 \text{ cm}^2$

21. a. $A = 34.9(12) + 52.5(20) = 1468.8 \text{ cm}^2$

b. Cost $= 330(\$0.008) = \$2.64$

25.

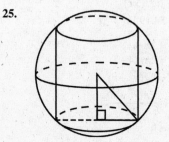

$h = d$ or $h = 2r$

In the triangle shown, $\frac{h}{2} = r$.

Then,

$$r^2 + r^2 = 6^2$$

$$2r^2 = 36$$

$$r^2 = 18$$

$$r = \sqrt{18} = 3\sqrt{2} \approx 4.24 \text{ in.}$$

$$h = 6\sqrt{2} \approx 8.49 \text{ in.}$$

29. a. $S = 4\pi r^2$

$S = 4\pi \cdot 3^2$

$S = 36\pi \approx 113.1 \text{ m}^2$

b. $V = \dfrac{4}{3}\pi r^3$

$V = \dfrac{4}{3}\pi \cdot 3^3$

$V = 36\pi \approx 113.1 \text{ m}^3$

33. $S = 4\pi r^2$

$S = 4\pi \cdot 3^2$

$S = 36\pi \approx 113.1 \text{ ft}^2$

The number of pints of paint needed to paint the

tank is $\dfrac{113.1}{40} = 2.836$ pints. 3 pints of paint

would have to be purchased.

37. a. Yes

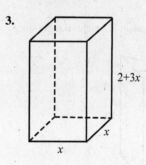

b. Yes

41. Congruent

45.

The solid of revolution looks like an inner tube.

CHAPTER REVIEW

1. $L = hP$

$L = 12(8 \cdot 7)$

$L = 672 \text{ in}^2$

2. $L = hP$

$L = 11(7 + 8 + 12)$

$L = 297 \text{ cm}^2$

3.

$2 + 3x$

x

x

$L = hP$

$480 = (2 + 3x) \cdot 4x$

$480 = 8x + 12x^2$

$0 = 12x^2 + 8x - 480$

$0 = 3x^2 + 2x - 120$

$0 = (3x + 20)(x - 6)$

$3x + 20 = 0 \quad$ or $\quad x - 6 = 0$

$x = -\dfrac{20}{3}$ or $\quad x = 6$; reject $x = -\dfrac{20}{3}$

Dimensions are 6 in. by 6 in. by 20 in.

$V = Bh$

$V = (6 \cdot 6)(20)$

$V = 720 \text{ in}^3$

4.

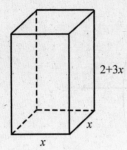

$2+3x$

x

x

$$L = 12(4x + 6)$$
$$360 = 48x + 72$$
$$48x = 288$$
$$x = 6$$

The dimensions of the box are $l = 9$ cm,
$w = 6$ cm, $h = 12$ cm.

$$T = L + 2B$$
$$T = 360 + 2(9 \cdot 6)$$
$$T = 360 + 108$$
$$T = 468 \text{ cm}^2$$

$$V = l \cdot w \cdot h$$
$$V = 9 \cdot 6 \cdot 12$$
$$V = 648 \text{ cm}^3$$

5. The base of the prism is a right triangle since
$$15^2 = 9^2 + 12^2$$

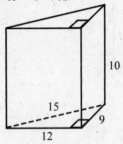

10

15

9

12

a. $L = hP$
$$L = 10(9 + 12 + 15)$$
$$L = 360 \text{ in}^2$$

b. The area of the base is
$$B = \frac{1}{2}bh$$
$$B = \frac{1}{2} \cdot 12 \cdot 9$$
$$B = 54 \text{ in}^2$$

$$T = L + 2B$$
$$T = 360 + 2 \cdot 54$$
$$T = 468 \text{ in}^2$$

c. $V = Bh$
$$V = 54 \cdot 10$$
$$V = 540 \text{ in}^3$$

6. a. $L = hP$
$$L = 13(6 \cdot 8)$$
$$L = 624 \text{ cm}^2$$

b. Using the 30°-60°-90° relationship, the apothem of the base is $4\sqrt{3}$.

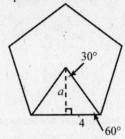

30°

a

4

60°

The area of the base is then equal to
$$B = \frac{1}{2}aP$$
$$B = \frac{1}{2} \cdot 4\sqrt{3} \cdot (6 \cdot 8)$$
$$B = 96\sqrt{3} \text{ cm}^2$$

$$T = L + 2B$$
$$T = 624 + 2 \cdot 96\sqrt{3}$$
$$T = 624 + 192\sqrt{3}$$
$$T \approx 956.55 \text{ cm}^2$$

c. $V = Bh$
$$V = \left(96\sqrt{3}\right) \cdot 13$$
$$V = 1248\sqrt{3} \approx 2161.6 \text{ cm}^3$$

7. $l^2 = a^2 + h^2$
$$l^2 = 5^2 + 8^2$$
$$l^2 = 89$$
$$l = \sqrt{89} \approx 9.43 \text{ cm}$$

8. $l^2 = a^2 + h^2$
$$12^2 = 9^2 + h^2$$
$$h^2 = 144 - 81$$
$$h^2 = 63$$
$$h = \sqrt{63} = 3\sqrt{7} \approx 7.94 \text{ in.}$$

9. $l^2 = r^2 + h^2$
$$l^2 = 5^2 + 7^2$$
$$l^2 = 74$$
$$l = \sqrt{74} \approx 8.60 \text{ in.}$$

10. $l^2 = r^2 + h^2$
$$(2r)^2 = r^2 + 6^2$$
$$4r^2 = r^2 + 36$$
$$3r^2 = 36$$
$$r^2 = 12$$
$$r = \sqrt{12} = 2\sqrt{3} \approx 3.46 \text{ cm}$$

11.

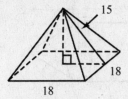

a. $L = \frac{1}{2}lP$

$L = \frac{1}{2} \cdot 15 \cdot (4 \cdot 18)$

$L = 540 \text{ in}^2$

b. $T = L + B$

$T = 540 + 18^2$

$T = 540 + 324$

$T = 864 \text{ in}^2$

c.

$l^2 = a^2 + h^2$

$15^2 = 9^2 + h^2$

$h^2 = 225 - 81$

$h^2 = 144$

$h = 12 \text{ in.}$

$V = \frac{1}{3}Bh$

$V = \frac{1}{3} \cdot 18^2 \cdot 12$

$V = 1296 \text{ in}^3$

12. a. Using the 30°-60°-90° relationship, the apothem is $a = 2\sqrt{3}$ cm.

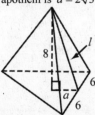

$l^2 = a^2 + h^2$

$l^2 = \left(2\sqrt{3}\right)^2 + 8^2$

$l^2 = 12 + 64$

$l^2 = 76$

$l = \sqrt{76} = 2\sqrt{19} \text{ cm}$

$L = \frac{1}{2}lP$

$L = \frac{1}{2} \cdot 2\sqrt{19} \cdot (12 \cdot 3)$

$L = 36\sqrt{19} = 156.92 \text{ cm}^2$

b. $T = L + B$

$T = 36\sqrt{19} + \frac{s^2\sqrt{3}}{4}$

$T = 36\sqrt{19} + \frac{12^2\sqrt{3}}{4}$

$T = 36\sqrt{19} + 36\sqrt{3} \approx 219.27 \text{ cm}^2$

c. $V = \frac{1}{3}Bh$

$V = \frac{1}{3}\left(36\sqrt{3}\right)(8)$

$V = 96\sqrt{3} \approx 166.28 \text{ cm}^3$

13. a. $L = hC$

$L = h \cdot 2\pi r$

$L = 10 \cdot 2\pi \cdot 6$

$L = 120\pi \text{ in}^2$

b. $T = L + 2B$

$T = 120\pi + 2\pi r^2$

$T = 120\pi + 2\pi \cdot 6^2$

$T = 120\pi + 72\pi$

$T = 192\pi \text{ in}^2$

c. $V = Bh$

$V = \pi r^2 h$

$V = \pi \cdot 6^2 \cdot 10$

$V = 360\pi \text{ in}^3$

14. a. $V = \frac{1}{2}Bh$

$V = \frac{1}{2}\pi r^2 h$

$V \approx \frac{1}{2} \cdot 3.14 \cdot 4^2 \cdot 14$

$V \approx 351.68 \text{ ft}^3$

b. The inside area and outside area represent the total area of the cylinder.

$T = L + 2B$

$T \approx 2(3.14)(4)(14) + 2(3.14) \cdot 4^2$

$T \approx 351.68 + 100.48$

$T \approx 452.16 \text{ ft}^2$

15. a.

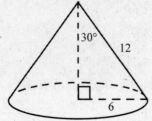

$$L = \frac{1}{2}lC$$
$$L = \frac{1}{2}l \cdot 2\pi r$$
$$L = l \cdot \pi r$$
$$L = 12 \cdot \pi \cdot 6$$
$$L = 72\pi \approx 226.19 \text{ cm}^2$$

b. $T = L + B$
$$T = 72\pi + \pi r^2$$
$$T = 72\pi + \pi \cdot 6^2$$
$$T = 72\pi + 36\pi$$
$$T = 108\pi \approx 339.29 \text{ cm}^2$$

c. $V = \frac{1}{3}Bh$
$$V = \frac{1}{3} \cdot \pi r^2 \cdot h$$
$$V = \frac{1}{3} \cdot \pi \cdot 6^2 \cdot 6\sqrt{3}$$
$$V = 72\pi\sqrt{3} \approx 391.78 \text{ cm}^3$$

16.

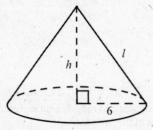

$$V = \frac{1}{3}Bh$$
$$V = \frac{1}{3}\pi r^2 \cdot h$$
$$96\pi = \frac{1}{3} \cdot \pi \cdot 6^2 \cdot h$$
$$96\pi = 12\pi h$$
$$h = \frac{96\pi}{12\pi} = 8$$

$l = 10$ using the Pythagorean Triple (6, 8, 10).

17. $S = 4\pi r^2$
$$S \approx 4 \cdot \frac{22}{7} \cdot 7^2$$
$$S \approx 616 \text{ in}^2$$

18. $V = \frac{4}{3}\pi r^3$
$$V \approx \frac{4}{3}(3.14) \cdot 6^3$$
$$V \approx 904.32 \text{ cm}^3$$

19.

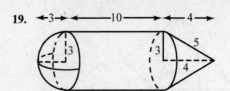

$$V = V_{\text{HEMISPHERE}} + V_{\text{CYLINDER}} + V_{\text{CONE}}$$
$$V = \frac{1}{2}\left(\frac{4}{3}\pi r^3\right) + Bh + \frac{1}{3}Bh$$
$$V = \frac{1}{2}\left(\frac{4}{3}\pi r^3\right) + \pi r^2 h + \frac{1}{3}\pi r^2 h$$
$$V = \frac{1}{2}\left(\frac{4}{3}\pi \cdot 3^3\right) + \pi \cdot 3^2 \cdot 10 + \frac{1}{3}\pi \cdot 3^2 \cdot 4$$
$$V = 18\pi + 90\pi + 12\pi$$
$$V = 120\pi \text{ units}^3$$

20. Let r equal the radius of the smaller sphere while $3r$ is the radius of the larger sphere.

$$\frac{\text{Surface Area of Smaller}}{\text{Surface Area of Larger}} = \frac{4\pi r^2}{4\pi(3r)^2} = \frac{r^2}{9r^2} = \frac{1}{2}$$

$$\frac{\text{Volume of Smaller}}{\text{Volume of Larger}} = \frac{\frac{4}{3}\pi r^3}{\frac{4}{3}\pi(3r)^3} = \frac{r^3}{27r^3} = \frac{1}{27}$$

21. The solid formed is a cone.
$$V = \frac{1}{3}Bh$$
$$V = \frac{1}{3} \cdot \pi r^2 \cdot h$$
$$V \approx \frac{1}{3} \cdot \frac{22}{7} \cdot 5^2 \cdot 7$$
$$V \approx 183\frac{1}{3} \text{ in}^3$$

22. The solid formed is a cylinder.
$$V = Bh$$
$$V = \pi r^2 h$$
$$V = \pi \cdot 6^2 \cdot 8$$
$$V = 288\pi \text{ cm}^3$$

23. The solid formed is a sphere.
$$V = \frac{4}{3}\pi r^3$$
$$V = \frac{4}{3}\pi \cdot 2^3$$
$$V = \frac{32\pi}{3} \text{ in}^3$$

24.

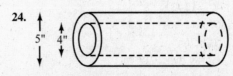

$$V = V_{\text{LARGER CYLINDER}} - V_{\text{SMALLER CYLINDER}}$$
$$V = \pi R^2 h - \pi r^2 h$$
$$V \approx (3.14)(5^2)(36) - (3.14)(4^2)(36)$$
$$V \approx 2826 - 1808.64$$
$$V \approx 1017.36 \text{ in}^3$$

25. $V = V_{CUBE} - V_{SPHERE}$

$V = e^3 - \dfrac{4}{3}\pi r^3$

$V = 14^3 - \dfrac{4}{3}\pi \cdot 7^3$

$V = 2744 - \dfrac{1372\pi}{3}$ in^3

26. a. An octahedron has eight faces which are equilateral triangles.

b. A tetrahedron has four faces which are equilateral triangles.

c. A dodecahedron has twelve faces which are regular pentagons.

27. $V = V_{CYLINDER} + V_{2\ HEMISPHERES} - V_{SPHERE}$

Because the volume of the 2 hemispheres equals the volume of a sphere, we have

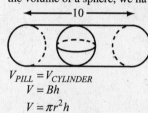

$V_{PILL} = V_{CYLINDER}$

$V = Bh$

$V = \pi r^2 h$

$V = \pi \cdot 2^2 \cdot 10$

$V = 40\pi$ mm^3

28. a. $V = 16$, $E = 24$, $F = 10$

$V + F = E + 2$

$16 + 10 = 24 + 2$

b. $V = 4$, $E = 6$, $F = 4$

$V + F = E + 2$

$4 + 4 = 6 + 2$

c. $V = 6$, $E = 12$, $F = 8$

$V + F = E + 2$

$6 + 8 = 12 + 2$

29. $V = 10 \cdot 3 \cdot 4 - 2 \cdot (1 \cdot 1 \cdot 3)$

$V = 120 - 6$

$V = 114$ in^3

30. a. $\dfrac{4}{8} = \dfrac{1}{2}$

b. $\dfrac{5}{8}$

31. a. $A = 6.5(12) = 78$ in^2

b. $A = 4\left(\dfrac{s^2\sqrt{3}}{4}\right) = 4\left(\dfrac{4^2\sqrt{3}}{4}\right) = 16\sqrt{3}$ cm^2

CHAPTER TEST

1. a. 15

b. 7

2. a. $P = 5 \cdot 3.2 = 16$ cm

$A = \dfrac{1}{2}aP$

$= \dfrac{1}{2}(2 \cdot 16)$

$= 16$ cm^2

b. $L = hP = 5 \cdot 16 = 80$

$T = L + 2B$

$T = 80 + 2\left(\dfrac{1}{2}aP\right)$

$T = 80 + 2(16)$

$T = 112$ cm^2

c. $V = Bh$

$V = \left(\dfrac{1}{2}aP\right)(5)$

$V = 16 \cdot 5$

$V = 80$ cm^3

3. a. 5

b. 4

4.

```
    |\
  6 | \ l
    |__\
     2
```

$6^2 = 2^2 + l^2$

$l^2 = 32$

$l = \sqrt{32} = 4\sqrt{2}$ ft

a. $L = \dfrac{1}{2}lP$

$L = \dfrac{1}{2}(4\sqrt{2})(16)$

$L = 32\sqrt{2}$ ft^2

b. $T = B + L$

$T = 16 + 32\sqrt{2} \approx 61.25$ ft^2

5.

```
    |\
 17 | \ l
    |__\
     8
```

$17^2 = 8^2 + l^2$

$l^2 = 225$

$l = 15$ ft

6.

$5^2 = 4^2 + l^2$

$l^2 = 9$

$l = 3$ in.

7. $V = \frac{1}{3}Bh$

$V = \frac{1}{3}(5 \cdot 5)(6)$

$V = 50$ ft^3

8. a. False

b. True

9. a. True

b. True

10. $V = 6$, $F = 8$

$V + F = E + 2$

$6 + 8 = E + 2$

$E = 12$

11.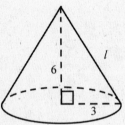

$l^2 = 3^2 + 6^2$

$l^2 = 45$

$l = \sqrt{45} = 3\sqrt{5}$ cm

12. a. $L = 2\pi rh$

$L = 2\pi \cdot 4 \cdot 6$

$L = 48\pi$ cm^2

b. $V = \pi r^2 h$

$V = \pi \cdot 4^2 \cdot 6$

$V = 96\pi$ cm^3

13.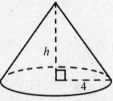

$V = \frac{1}{3}\pi r^2 h$

$32\pi = \frac{1}{3}\pi \cdot 4^2 \cdot h$

$h = 6$ in.

14. a. $\frac{4}{8} = \frac{1}{2}$

b. $\frac{3}{8}$

15. a. $S = 4\pi r^2$

$S = 4\pi \cdot 10^2$

$S = 400\pi \approx 1256.6$ ft^2

b. $V = \frac{4}{3}\pi r^3$

$V = \frac{4}{3}\pi \cdot 10^3$

$V = \frac{4000}{3}\pi \approx 4188.8$ ft^3

16. $V = \frac{4}{3}\pi r^3$

$V = \frac{4}{3}\pi(10)^3$

$V = \frac{4000}{3}\pi$ ft^3

Volume = Rate · Time

Time = Volume ÷ Rate

Time $= \frac{4000}{3}\pi$ ft^3 $\div 8\pi \dfrac{ft^3}{\min}$

Time $= \frac{4000}{3}\pi$ ft^3 $\cdot \dfrac{1}{8\pi} \dfrac{\min}{ft^3}$

Time =

$\frac{500}{3}$ min $= 166\frac{2}{3}$ or 2 hours 47 minutes

Chapter 10: Analytic Geometry

SECTION 10.1: The Rectangular Coordinate System

1.

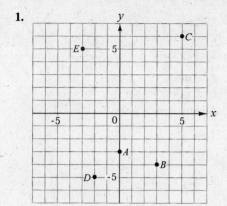

5. The segment is horizontal so $d = 7 - b$ if $7 > b$
$\therefore 7 - b = 3.5$
$-b = -3.5$
$b = 3.5$
If $b > 7$, then $d = b - 7$.
$\therefore b - 7 = 3.5$
$b = 10.5$

9. a. $M = \left(\dfrac{x_1 + x_2}{2}, \dfrac{y_1 + y_2}{2} \right)$
$M = \left(\dfrac{0+4}{2}, \dfrac{(-3)+0}{2} \right)$
$M = \left(\dfrac{4}{2}, -\dfrac{3}{2} \right) = \left(2, -\dfrac{3}{2} \right)$

b. $M = \left(\dfrac{(-2)+4}{2}, \dfrac{5+(-3)}{2} \right)$
$M = \left(\dfrac{2}{2}, \dfrac{3}{2} \right)$
$M = (1, 1)$

c. $M = \left(\dfrac{3+5}{2}, \dfrac{2+(-2)}{2} \right)$
$M = \left(\dfrac{8}{2}, \dfrac{0}{2} \right)$
$M = (4, 0)$

d. $M = \left(\dfrac{a+0}{2}, \dfrac{0+b}{2} \right)$
$M = \left(\dfrac{a}{2}, \dfrac{b}{2} \right)$

13. a. $\left(4, -\dfrac{5}{2} \right)$

b. $(0, 4)$

c. $\left(\dfrac{7}{2}, -1 \right)$

d. (a, b)

17. a. $(-3, -4)$

b. $(-2, 0)$

c. $(-a, 0)$

d. $(-b, c)$

21. $M = \left(\dfrac{x_1 + x_2}{2}, \dfrac{y_1 + y_2}{2} \right)$
$(2.1, -5.7) = \left(\dfrac{x+1.7}{2}, \dfrac{y+2.3}{2} \right)$
$\dfrac{x+1.7}{2} = 2.1 \quad \text{and} \quad \dfrac{y+2.3}{2} = -5.7$
$x + 1.7 = 4.2 \quad \text{and} \quad y + 2.3 = -11.4$
$x = 2.5 \quad \text{and} \quad y = -13.7$
$B = (2.5, -13.7)$

25. a. $AB = 4 - 0 = 4$
$BC = \sqrt{(2-4)^2 + (5-0)^2}$
$= \sqrt{(-2)^2 + 5^2}$
$= \sqrt{4 + 25} = \sqrt{29}$
$AC = \sqrt{(2-0)^2 + (5-0)^2}$
$= \sqrt{2^2 + 5^2}$
$= \sqrt{4 + 25} = \sqrt{29}$
Because $BC = AC$, $\triangle ABC$ is isosceles.

b. $DE = 4 - 0 = 4$
$DF = \sqrt{(2-0)^2 + (2\sqrt{3} - 0)^2}$
$= \sqrt{2^2 + (2\sqrt{3})^2}$
$= \sqrt{4 + 12} = \sqrt{16} = 4$
$EF = \sqrt{(2-4)^2 + (2\sqrt{3} - 0)^2}$
$= \sqrt{(-2)^2 + (2\sqrt{3})^2}$
$= \sqrt{4 + 12} = \sqrt{16} = 4$
Because $DE = DF = EF$, $\triangle DEF$ is equilateral.

c. $GH = \sqrt{(-2-[-5])^2 + (6-2)^2}$

$\quad\quad = \sqrt{3^2 + 4^2}$

$\quad\quad = \sqrt{9+16} = \sqrt{25} = 5$

$\quad GK = \sqrt{(2-[-5])^2 + (3-2)^2}$

$\quad\quad = \sqrt{7^2 + 1^2}$

$\quad\quad = \sqrt{49+1} = \sqrt{50} = 5\sqrt{2}$

$\quad HK = \sqrt{(2-[-2])^2 + (3-6)^2}$

$\quad\quad = \sqrt{4^2 + (-3)^2}$

$\quad\quad = \sqrt{16+9} = \sqrt{25} = 5$

Because $GH = HK$ and

$(GH)^2 + (HK)^2 = (GK)^2$, $\triangle GHK$ is an isosceles right triangle.

29. Call the third vertex (a, b). Because the three sides must be of the same length and $AB = 2a$, we have

$$\sqrt{(a-0)^2 + (b-0)^2} = 2a$$
$$\sqrt{a^2 + b^2} = 2a$$

Squaring, $a^2 + b^2 = 4a^2$

$$b^2 = 3a^2$$
$$b = \pm\sqrt{3a^2}$$
$$b = \pm a\sqrt{3}$$

The third vertex is $(a, a\sqrt{3})$ or $(a, -a\sqrt{3})$.

33. $A_{MNQ} = A_{\text{RECT}} - (A_{\triangle 1} + A_{\triangle 2} + A_{\triangle 3})$

$\quad A = (7)(5) - \left[\dfrac{1}{2}(1)(5) + \dfrac{1}{2}(6)(4) + \dfrac{1}{2}(7)(1)\right]$

$\quad A = 35 - \left[\dfrac{5}{2} + 12 + \dfrac{7}{2}\right]$

$\quad A = 35 - 18$

$\quad A = 17$

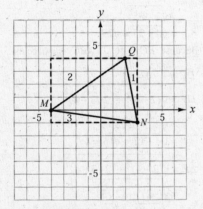

37. a. A cone is formed; $r = 9, h = 5$

$\quad V = \dfrac{1}{3}\pi r^2 h$

$\quad V = \dfrac{1}{3}\pi \cdot 9^2 \cdot 5$

$\quad V = 135\pi$ units^2

b. A cone is formed; $r = 5, h = 9$

$\quad V = \dfrac{1}{3}\pi r^2 h$

$\quad V = \dfrac{1}{3}\pi \cdot 5^2 \cdot 9$

$\quad V = 75\pi$ units^2

41. a. $L = 2\pi r h$

$\quad L = 2\pi \cdot 5 \cdot 9$

$\quad L = 90\pi$ units^2

b. $L = 2\pi r h$

$\quad L = 2\pi \cdot 9 \cdot 5$

$\quad L = 90\pi$ units^2

47. a. $(5, 4)$

b. $(1, 8)$

c. $(3, 2)$

SECTION 10.2: Graphs of Linear Equations and Slope

1. $3x + 4y = 12$ has intercepts $(4, 0)$ and $(0, 3)$.

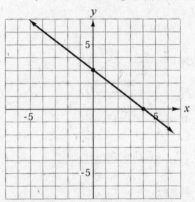

5. $2x + 6 = 0$ is equivalent to $x = -3$. It is a vertical line with x-intercept $(-3, 0)$.

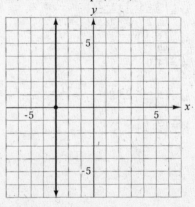

9. a. $m = \dfrac{y_2 - y_1}{x_2 - x_1}$

$m = \dfrac{5 - (-3)}{4 - 2}$

$m = \dfrac{8}{2}$

$m = 4$

b. $m = \dfrac{7 - (-2)}{3 - 3} = \dfrac{9}{0}$

m is undefined.

c. $m = \dfrac{-2 - (-1)}{2 - 1}$

$m = \dfrac{-1}{1}$

$m = -1$

d. $m = \dfrac{5 - 5}{(-1.3) - (-2.7)}$

$m = \dfrac{0}{1.4}$

$m = 0$

e. $m = \dfrac{d - b}{c - a}$

f. $m = \dfrac{b - 0}{0 - a}$

$m = -\dfrac{b}{a}$

13. a. $m_{\overline{AB}} = \dfrac{2 - 5}{0 - (-2)}$

$m_{\overline{AB}} = -\dfrac{3}{2}$

$m_{\overline{BC}} = \dfrac{-4 - 2}{4 - 0}$

$m_{\overline{BC}} = \dfrac{-6}{4}$

$m_{\overline{BC}} = -\dfrac{3}{2}$

Because $m_{\overline{AB}} = m_{\overline{BC}}$, the points A, B and C are collinear.

b. $m_{\overline{DE}} = \dfrac{-2 - (-1)}{2 - (-1)}$

$m_{\overline{DE}} = -\dfrac{1}{3}$

$m_{\overline{EF}} = \dfrac{-5 - (-2)}{5 - 2}$

$m_{\overline{EF}} = \dfrac{-3}{3}$

$m_{\overline{EF}} = -1$

Because $m_{\overline{DE}} \neq m_{\overline{EF}}$, the points D, E and F are noncollinear.

17. a. 2

b. $-\dfrac{4}{3}$

c. $-\dfrac{1}{3}$

d. $-\dfrac{h + j}{f + g}$

21. $2x + 3y = 6$ contains $(3, 0)$ and $(0, 2)$ so that its slope is $m_1 = \dfrac{2 - 0}{0 - 3} = -\dfrac{2}{3}$.

$3x - 2y = 12$ contains $(4, 0)$ and $(0, -6)$ so that its slope is $m_2 = \dfrac{-6 - 0}{0 - 4} = \dfrac{-6}{-4} = \dfrac{3}{2}$.

Because $m_1 \cdot m_2 = -1$, these lines are perpendicular.

25. The 2 lines are perpendicular if $m_1 \cdot m_2 = -1$.

$m_1 = \dfrac{2 - (-3)}{3 - 2} = \dfrac{5}{1} = 5$

$m_2 = \dfrac{-1 - 4}{x - (-2)} = \dfrac{-5}{x + 2}$

$\dfrac{5}{1} \cdot \dfrac{-5}{x + 2} = -1$

$\dfrac{-25}{x + 2} = -1$

$-25 = -1(x + 2)$

$-25 = -1x - 2$

$-23 = -1x$

$x = 23$

29. First plot $(0, 5)$. If $m = -\dfrac{3}{4}$, then $m = \dfrac{-3}{4}$; a change in y of -3 corresponds to a change in x of 4.

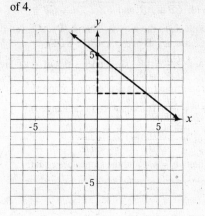

33. Let $A = (6, 5)$, $B = (-3, 0)$, and $C = (4, -2)$.

$m_{\overline{AB}} = \dfrac{0 - 5}{-3 - 6} = \dfrac{-5}{-9} = \dfrac{5}{9}$

$m_{\overline{BC}} = \dfrac{-2 - 0}{4 - (-3)} = \dfrac{-2}{7}$

$m_{\overline{AC}} = \dfrac{-2 - 5}{4 - 6} = \dfrac{-7}{-2} = \dfrac{7}{2}$

Because $m_{\overline{BC}} \cdot m_{\overline{AC}} = -1$, $\overline{BC} \perp \overline{AC}$ and $\triangle ABC$ is a right triangle.

37. $m_{\overline{VT}} = \dfrac{e-e}{(c-d)-(a+d)} = \dfrac{0}{c-a-2d} = 0$

$m_{\overline{RS}} = \dfrac{b-b}{c-a} = \dfrac{0}{c-a} = 0$

$\therefore \overline{VT} \parallel \overline{RS}$

$RV = \sqrt{[(a+d)-a]^2 + (e-b)^2}$

$\quad = \sqrt{d^2 + (e-b)^2}$

$\quad = \sqrt{d^2 + e^2 - 2be + b^2}$

$ST = \sqrt{[c-(c-d)]^2 + (b-e)^2}$

$\quad = \sqrt{d^2 + (b-e)^2}$

$\quad = \sqrt{d^2 + b^2 - 2be + e^2}$

$\therefore \overline{RV} \parallel \overline{ST}$

$RSTV$ is an isosceles trapezoid.

41. If $l_1 \parallel l_2$, then $\angle A \cong \angle D$.

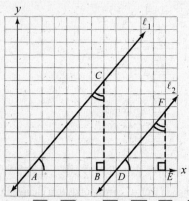

With $\overline{CB} \perp \overline{AB}$ and $\overline{FE} \perp \overline{DE}$, $\overline{CB} \parallel \overline{FE}$

(both $\perp$ to the x-axis). Then $\angle B \cong \angle E$ (all rt. $\angle s$ are $\cong$). By AA, $\triangle ABC \sim \triangle DEF$.

Then $\dfrac{CB}{FE} = \dfrac{AB}{DE}$ since corr. sides of $\sim \triangle s$ are proportional. By a property of proportions,

$\dfrac{CB}{AB} = \dfrac{FE}{DE}$ (means were interchanged).

But $m_1 = \dfrac{CB}{AB}$ and $m_2 = \dfrac{FE}{DE}$.

Then $m_1 = m_2$ and the slopes are equal.

SECTION 10.3: Preparing to Do Analytic Proofs

1. a. $d = \sqrt{(x_2 - x_1)^2 + (y_2 - y_1)^2}$

$d = \sqrt{(0-a)^2 + (a-0)^2}$

$d = \sqrt{(-a)^2 + a^2}$

$d = \sqrt{a^2 + a^2}$

$d = \sqrt{2a^2} = a\sqrt{2}$ if $a > 0$

b. $m = \dfrac{y_2 - y_1}{x_2 - x_1}$

$m = \dfrac{d-b}{c-a}$

5. $\overline{AB}$ is horizontal and $\overline{BC}$ is vertical. $\therefore \overline{AB} \perp \overline{BC}$. Hence $\angle B$ is a right $\angle$ and $\triangle ABC$ is a right triangle.

9. $m_{\overline{MN}} = 0$ and $m_{\overline{QP}} = 0$. $\therefore \overline{MN} \parallel \overline{QP}$

$\overline{QM}$ and $\overline{PN}$ are both vertical;

$\therefore \overline{QM} \parallel \overline{PN}$. Hence, $MQPN$ is a parallelogram.

Since $\overline{QM}$ is vertical and $\overline{MN}$ is horizontal, $\angle QMN$ is a right angle. Because parallelogram $MQPN$ has a right $\angle$, it is also a rectangle.

13. $M = (0, 0); N = (r, 0); P = (r + s, t)$

17. a. Square

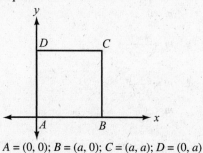

$A = (0, 0); B = (a, 0); C = (a, a); D = (0, a)$

b. Square (with midpoints of sides)
$A = (0, 0); B = (2a, 0)$
$C = (2a, 2a); D = (0, 2a)$

21. a. Isosceles triangle

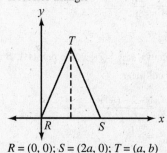

$R = (0, 0); S = (2a, 0); T = (a, b)$

b. Isosceles triangle (with midpoints)
$R = (0, 0); S = (4a, 0); T = (2a, 2b)$

25. Parallelogram $ABCD$ with $\overline{AC} \perp \overline{DB}$

$$m_{\overline{AC}} = \frac{c-0}{a+b-0} = \frac{c}{a+b}$$

$$m_{\overline{DB}} = \frac{0-c}{a-b} = \frac{-c}{a-b}$$

$$\therefore \frac{c}{a+b} \cdot \frac{-c}{a-b} = -1$$

$$\frac{-c^2}{a^2-b^2} = -1$$

$$-c^2 = -(a^2-b^2)$$

$$c^2 = a^2 - b^2$$

29. a. a is positive

b. $-a$ is negative

c. $AB = a - (-a) = 2a$

33. The line segment joining the midpoints of the two nonparallel sides of a trapezoid is parallel to the bases of the trapezoid.

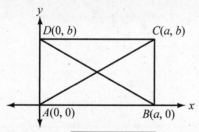

SECTION 10.4: Analytic Proofs

1. The diagonals of a rectangle are equal in length.
Proof: Let rectangle $ABCD$ have vertices as shown.

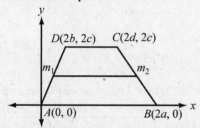

Then $AC = \sqrt{(a-0)^2 + (b-0)^2}$
$= \sqrt{a^2 + b^2}$

Also $DB = \sqrt{(a-0)^2 + (0-b)^2}$
$= \sqrt{a^2 + (-b)^2} = \sqrt{a^2 + b^2}$

Then $AC = DB$, and the diagonals of the rectangle are equal in length.

5. The median from the vertex of an isosceles triangle to the base is perpendicular to the base.

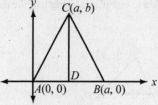

Proof: The triangle ABC with vertices as shown is isosceles because $AC = BC$.

Let D be the midpoint of $\overline{AB}$.

$$D = \left(\frac{0+2a}{2}, \frac{0+0}{2}\right) = (a, 0).$$

$m_{\overline{AB}} = \frac{0-0}{2a-0} = 0$; that is $\overline{AB}$ is horizontal.

$m_{\overline{CD}} = \frac{b-0}{a-a} = \frac{b}{0}$ which is undefined; that is $\overline{CD}$ is vertical.

Then $\overline{CD} \perp \overline{AB}$ and the median to the base of the isosceles triangle is perpendicular to the base.

9. The segments which join the midpoints of the consecutive sides of a rectangle form a rhombus.

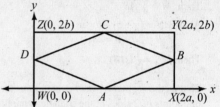

Proof: With vertices as shown, $WXYZ$ is a rectangle.
Midpoints of the sides of the rect. Are

$$A = \left(\frac{0+2a}{2}, \frac{0+0}{2}\right) = (a, 0)$$

$$B = \left(\frac{2a+2a}{2}, \frac{0+2b}{2}\right) = (2a, b)$$

$$C = \left(\frac{0+2a}{2}, \frac{2b+2b}{2}\right) = (a, 2b)$$

$$D = \left(\frac{0+0}{2}, \frac{0+2b}{2}\right) = (0, b)$$

$m_{\overline{AB}} = \frac{b-0}{2a-a} = \frac{b}{a}$ and

$m_{\overline{DC}} = \frac{b-a}{0-a} = \frac{-b}{-a} = \frac{b}{a}$, so $\overline{AB} \parallel \overline{DC}$.

$m_{\overline{DA}} = \frac{b-0}{0-a} = \frac{b}{-a}$ and

$m_{\overline{CB}} = \frac{2b-b}{a-2a} = \frac{-b}{-a} = \frac{b}{a}$, so $\overline{DA} \parallel \overline{CB}$.

Then $ABCD$ is a parallelogram.
We need to show that 2 adjacent sides are congruent (equal in length).

$$AB = \sqrt{(2a-a)^2 + (b-0)^2} = \sqrt{a^2 + b^2}$$

$$BC = \sqrt{(a-2a)^2 + (2b-b)^2}$$

$$= \sqrt{(-a)^2 + b^2} = \sqrt{a^2 + b^2}$$

13. The segment which joins the midpoints of 2 sides of a triangle is parallel to the third side and has a length equal to one-half the length of the third side.

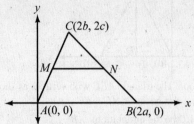

Proof: Let $\triangle ABC$ have vertices as shown. With M and N the midpoints of $\overline{AC}$ and $\overline{BC}$ respectively,

$$M = \left(\frac{0+2b}{2}, \frac{0+2c}{2}\right) = (b, c) \text{ and}$$

$$N = \left(\frac{2a+2b}{2}, \frac{0+2c}{2}\right) = (a+b, c).$$

Now $m_{\overline{MN}} = \frac{c-c}{(a+b)-b} = \frac{0}{a} = 0$ and

$$m_{\overline{AB}} = \frac{0-0}{2a-0} = \frac{0}{2a} = 0$$

Then $\overline{MN} \parallel \overline{AB}$.

Also $MN = (a+b) - b = a$

$AB = 2a - 0 = 2a$

Then $MN = \frac{1}{2}(AB)$

That is, the segment $\left(\overline{MN}\right)$ which joins the midpoints of 2 sides of the triangle is parallel to the third side and equals one-half its length.

17. If the diagonals of a parallelogram are perpendicular, then the parallelogram is a rhombus.

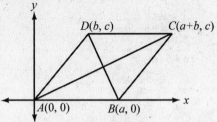

Proof: Let parallelogram $ABCD$ have vertices as shown. If $\overline{AC} \perp \overline{DB}$, then $m_{\overline{AC}} \cdot m_{\overline{DB}} = -1$.

Because $m_{\overline{AC}} = \frac{c-0}{(a+b)-0} = \frac{c}{a+b}$ and

$m_{\overline{DB}} = \frac{0-c}{a-b} = \frac{-c}{a-b}$, it follows that

$$\frac{c}{a+b} \cdot \frac{-c}{a-b} = -1$$

$$\frac{-c^2}{a^2-b^2} = -1$$

$$-c^2 = -1(a^2 - b^2)$$

$$-c^2 = -a^2 + b^2$$

$$a^2 = b^2 + c^2 \, (*)$$

For $ABCD$ to be a rhombus, we must show that two adjacent sides are congruent. $AB = a - 0 = a$

and $AD = \sqrt{(b-0)^2 + (c-0)^2} = \sqrt{b^2 + c^2}$.

Because $a^2 = b^2 + c^2 \, (*)$, it follows that

$a = \sqrt{b^2 + c^2}$ and $AB = AD$. Then parallelogram $ABCD$ is a rhombus.

21. $Ax + By = C$

$By = -Ax + C$

$$y = -\frac{A}{B}x + \frac{C}{B}$$

So $m_1 = -\frac{A}{B}$

$Bx - Ay = D$

$-Ay = -Bx + D$

$$y = \frac{B}{A}x - \frac{D}{A}$$

So $m_2 = \frac{B}{A}$

Since $m_1 \cdot m_2 = -\frac{A}{B} \cdot \frac{B}{A} = -1$, then the lines are

perpendicular.

25. $x^2 + y^2 = r^2$

$x^2 + y^2 = 3^2$

$x^2 + y^2 = 9$

SECTION 10.5: Equations of Lines

1. Dividing by 8, $8x + 16y = 48$ becomes
$x + 2y = 6$.
Then $2y = -1x + 6$
$$\frac{1}{2}(2y) = \frac{1}{2}(-1x + 6)$$
$$y = -\frac{1}{2}x + 3$$

5. $y = 2x - 3$ has $m = 2$ and $b = -3$. Plot the point $(0, -3)$. Then draw the line for which an increase of 2 in y corresponds to an increase of 1 in x.

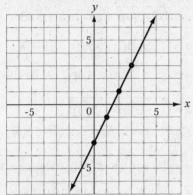

9. Using $y = mx + b$,
$$y = -\frac{2}{3}x + 5$$
$$3y = -2x + 15$$
$$2x + 3y = 15$$

13. $m = \frac{1 - (-1)}{3 - 0} = \frac{2}{3}$

Using $y = mx + b$, $y = \frac{2}{3}x - 1$.

Mult. by 3,
$$3y = 2x - 3$$
$$-2x + 3y = -3$$

17. The graph contains $(2, 0)$ and $(0, -2)$.
$$m = \frac{-2 - 0}{0 - 2} = \frac{-2}{-2} = 1.$$
Using $y = mx + b$,
$$y = 1x + 2$$
$$-x + y = -2$$

21. The line $y = \frac{3}{4}x - 5$ has slope $m_1 = \frac{3}{4}$. The desired line has $m_2 = -\frac{4}{3}$. Using $y = mx + b$,

$y = -\frac{4}{3}x - 4$. Mult. by 3,
$$3y = -4x - 12$$
$$4x + 3y = -12$$

25. Perpendicular slope is $-\frac{b}{a}$.
$$y - h = -\frac{b}{a}(x - g)$$
$$y = h = -\frac{b}{a}x + \frac{bg}{a}$$
$$y = -\frac{b}{a}x + \frac{bg + ha}{a}$$

29. $2x + y = 6$
$\quad y = -2x + 6$
$3x - y = 19$
$\quad y = 3x - 19$

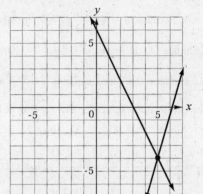

The lines intersect at $(5, -4)$.

33. $2x + y = 8$
$\underline{3x - y = 7}$
$\quad 5x = 15$
$\quad\quad x = 3$
$$2(3) + y = 8$$
$$6 + y = 8$$
$$y = 2$$

37. $2x + 3y = 4$ Multiply by -3
$3x - 4y = 23$ Multiply by 2

$-6x - 9y = -12$
$\underline{6x - 8y = 46}$
$\quad -17y = 34$
$\quad\quad\quad y = -2$
$$2x + 3(-2) = 4$$
$$2x - 6 = 4$$
$$2x = 10$$
$$x = 5$$
The point of intersection is $(5, -2)$.

41. $bx + c = a$
$\quad bx = a - c$
$$x = \frac{a - c}{b}$$
and $y = a$.
$$\left(\frac{a - c}{b}, a\right)$$

45. The altitudes of a triangle are concurrent.

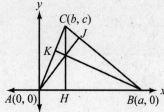

Proof: For $\triangle ABC$, let $\overline{CH}$, $\overline{AJ}$, and $\overline{BK}$ name the altitudes. Because $\overline{AB}$ is horizontal $\left(m_{\overline{AB}} = 0\right)$, $\overline{CH}$ is vertical and has the equation $x = b$. Because $m_{\overline{BC}} = \dfrac{c-0}{b-a} = \dfrac{c}{b-a}$, the slope of altitude $\overline{AJ}$ is $m_{\overline{AJ}} = -\dfrac{b-a}{c} = \dfrac{a-b}{c}$. Since $\overline{AJ}$ contains (0, 0), its equation is $y = \dfrac{a-b}{c}x$.

The intersection of altitudes $\overline{CH}$ ($x = b$) and $\overline{AJ}$ $\left(y = \dfrac{a-b}{c}x\right)$ is at $x = b$ so that

$$y = \frac{a-b}{c} \cdot b = \frac{b(a-b)}{c} = \frac{ab-b^2}{c}.$$ That is, $\overline{CH}$ and $\overline{AJ}$ intersect at $\left(b, \dfrac{ab-b^2}{c}\right)$. The remaining altitude is $\overline{BK}$. Since $m_{\overline{AC}} = \dfrac{c-0}{b-0} = \dfrac{c}{b}$,

$m_{\overline{BK}} = -\dfrac{b}{c}$. Because $\overline{BK}$ contains $(a, 0)$, its equation is $y - 0 = -\dfrac{b}{c}(x-a)$ or $y = \dfrac{-b}{c}(x-a)$.

For the three altitudes to be concurrent,

$\left(b, \dfrac{ab-b^2}{c}\right)$ must lie on the line $y = \dfrac{-b}{c}(x-a)$.

Substitution leads to

$$\frac{ab-b^2}{c} = \frac{-b}{c}(b-a)$$
$$\frac{ab-b^2}{c} = \frac{-b(b-a)}{c}$$
$$\frac{ab-b^2}{c} = \frac{-b^2+ab}{c}, \text{ which is true.}$$

Therefore, the three altitudes are concurrent.

CHAPTER REVIEW

1. a. $d = y_2 - y_1$ (if $y_2 > y_1$) $= 4 - (-3) = 7$

   b. $d = x_2 - x_1$ (if $x_2 > x_1$) $= 1 - (-5) = 6$

   c. $d = \sqrt{(x_2 - x_1)^2 + (y_2 - y_1)^2}$

      $d = \sqrt{(7-(-5))^2 + (-3-2)^2}$

      $d = \sqrt{12^2 + (-5)^2} = \sqrt{144+25} = \sqrt{169} = 13$

   d. $d = \sqrt{(x-(x-3))^2 + ((y-2)-(y+2))^2}$

      $d = \sqrt{3^2 + (-4)^2} = \sqrt{9+16} = \sqrt{25} = 5$

2. a. $d = y_2 - y_1$ (if $y_2 > y_1$) $= 5 - (-3) = 8$

   b. $d = x_2 - x_1$ (if $x_2 > x_1$) $= 3 - (-7) = 10$

   c. $d = \sqrt{(4-(-4))^2 + (5-1)^2}$

      $d = \sqrt{8^2 + 4^2} = \sqrt{64+16} = \sqrt{80} = 4\sqrt{5}$

   d. $d = \sqrt{((x+4)-(x-2))^2 + ((y+5)-(y-3))^2}$

      $d = \sqrt{6^2 + 8^2} = \sqrt{36+64} = \sqrt{100} = 10$

3. a. $M = \left(\dfrac{x_1 + x_2}{2}, \dfrac{y_1 + y_2}{2}\right)$

      $M = \left(\dfrac{6+6}{2}, \dfrac{4+(-3)}{2}\right) = \left(6, \dfrac{1}{2}\right)$

   b. $M = \left(\dfrac{1+(-5)}{2}, \dfrac{4+4}{2}\right) = (-2, 4)$

   c. $M = \left(\dfrac{(-5)+7}{2}, \dfrac{2+(-3)}{2}\right) = \left(1, -\dfrac{1}{2}\right)$

   d. $M = \left(\dfrac{(x-3)+x}{2}, \dfrac{(y+2)+(y-2)}{2}\right)$

      $M = \left(\dfrac{2x-3}{2}, y\right)$

4. a. $M = \left(\dfrac{2+2}{2}, \dfrac{(-3)+5}{2}\right) = (2, 1)$

   b. $M = \left(\dfrac{3+(-7)}{2}, \dfrac{(-2)+(-2)}{2}\right) = (-2, -2)$

   c. $M = \left(\dfrac{(-4)+4}{2}, \dfrac{1+5}{2}\right) = (0, 3)$

   d. $M = \left(\dfrac{(x-2)+(x+4)}{2}, \dfrac{(y-3)+(y+5)}{2}\right)$

      $M = (x+1, y+1)$

5. **a.** $m = \dfrac{y_2 - y_1}{x_2 - x_1}$

$m = \dfrac{(-3) - 4}{6 - 6} = \dfrac{-7}{0}$ m is undefined.

b. $m = \dfrac{4 - 4}{-5 - 1} = \dfrac{0}{-6} = 0$

c. $m = \dfrac{-3 - 2}{7 - (-5)} = \dfrac{-5}{12}$

d. $m = \dfrac{(y - 2) - (y + 2)}{x - (x - 3)} = \dfrac{-4}{3}$

6. **a.** $m = \dfrac{5 - (-3)}{2 - 2} = \dfrac{8}{0}$ m is undefined.

b. $m = \dfrac{-2 - (-2)}{-7 - 3} = \dfrac{0}{-10} = 0$

c. $m = \dfrac{5 - 1}{4 - (-4)} = \dfrac{4}{8} = \dfrac{1}{2}$

d. $m = \dfrac{(y + 5) - (y - 3)}{(x + 4) - (x - 2)} = \dfrac{8}{6} = \dfrac{4}{3}$

7. $M = \left(\dfrac{x_1 + x_2}{2}, \dfrac{y_1 + y_2}{2} \right)$

$(2, 1) = \left(\dfrac{8 + x}{2}, \dfrac{10 + y}{2} \right)$

$\dfrac{8 + x}{2} = 2$ and $\dfrac{10 + y}{2} = 1$

$8 + x = 4$ and $10 + y = 2$

$x = -4$ and $y = -8$

$B = (-4, \ -8)$

8. Due to symmetry, $R = (3, 7)$.

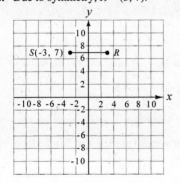

9. $m = \dfrac{y_2 - y_1}{x_2 - x_1}$

$-3 = \dfrac{3 - 1}{x - 2}$

$-3 = \dfrac{2}{x - 2}$

$-3(x - 2) = 2$

$-3x + 6 = 2$

$-3x = -4$

$x = \dfrac{4}{3}$

10. $m = \dfrac{y_2 - y_1}{x_2 - x_1}$

$\dfrac{-6}{7} = \dfrac{y - 2}{2 - (-5)}$

$\dfrac{-6}{7} = \dfrac{y - 2}{7}$

$-6 = y - 2$

$y = -4$

11. **a.** $x + 3y = 6$

$y = -\dfrac{1}{3}x + 2$

$m = -\dfrac{1}{3}$

$3x - y = -7$

$y = 3x + 7$

$m = 3$

Since $m_1 \cdot m_2 = -1$, the lines are perpendicular.

b. $2x - y = -3$

$y = 2x + 3$

$m = 2$ and $b = 3$

$y = 2x - 14$

$m = 2$ and $b = -14$

Since $m_1 = m_2$, the lines are parallel.

c. $y + 2 = -3(x - 5)$

$y = -3x + 13$

$m = -3$

$2y = 6x + 11$

$y = 3x + \dfrac{11}{2}$

$m = 3$

The lines are neither parallel nor perpendicular.

d. $0.5x + y = 0$

$y = -\dfrac{1}{2}x$

$m = -\dfrac{1}{2}$

$2x - y = 10$

$y = 2x - 10$

$m = 2$

Since $m_1 \cdot m_2 = -1$, the lines are perpendicular.

12. Let $A = (-6, 5)$, $B = (1, 7)$, and $C = (16, 10)$.
The points would be collinear if
$$m_{\overline{AB}} = m_{\overline{BC}}$$
$$m_{\overline{AB}} = \frac{7-5}{1-(-6)} = \frac{2}{7}$$
$$m_{\overline{BC}} = \frac{10-7}{16-1} = \frac{3}{15} = \frac{1}{5}$$
The points are not collinear.

13. Let $A = (-2, 3)$, $B = (x, 6)$, and $C = (8, 8)$. If
$m_{\overline{AB}} = m_{\overline{BC}}$, then A, B, and C are collinear.
$$m_{\overline{AB}} = \frac{6-3}{x-(-2)} = \frac{3}{x+2}$$
$$m_{\overline{BC}} = \frac{8-x}{8-x} = \frac{2}{8-x}$$
$$\frac{3}{x+2} = \frac{2}{8-x}$$
$$3(8-x) = 2(x+2)$$
$$24 - 3x = 2x + 4$$
$$20 = 5x$$
$$x = 4$$

14. Intercepts are $(7, 0)$ and $(0, 3)$.

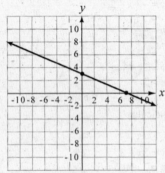

15. $4x - 3y = 9$
$$-3y = -4x + 9$$
$$y = \frac{4}{3}x - 3$$

First, plot the point $(0, -3)$. Then locate a second point for which an increase of 4 in y corresponds to an increase of 3 in x.

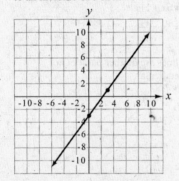

16. $y + 2 = \frac{-2}{3}(x - 1)$

$y - (-2) = \frac{-2}{3}(x - 1)$ is a line which contains

$(1, -2)$ and has slope $m = \frac{-2}{3}$. First plot

$(1, -2)$. Then from that point draw a line which

has $m = \frac{-2}{3}$.

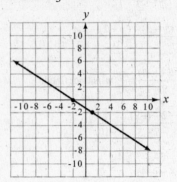

17. a. $m = \frac{6-3}{-3-2} = \frac{3}{-5}$ or $\frac{-3}{5}$
$$y - 3 = \frac{-3}{5}(x - 2)$$
$$5(y - 3) = -3(x - 2)$$
$$5y - 15 = -3x + 6$$
$$3x + 5y = 21$$

b. $m = \frac{-9-(-3)}{8-6} = \frac{-6}{2} = -3$
$$y - (-1) = -3(x - (-2))$$
$$y + 1 = -3(x + 2)$$
$$y + 1 = -3x - 6$$
$$3x + y = -7$$

c. $x + 2y = 4$
$$y = \frac{-1}{2}x + 2$$

Since $m_1 = \frac{-1}{2}$, the desired line has $m_2 = 2$.
$$y - (-2) = 2(x - 3)$$
$$y + 2 = 2x + 6$$
$$-2x + y = -8$$

d. A line parallel to the x-axis has the form
$y = b$. $\therefore y = 5$

18. Let $A = (-2, -3)$, $B = (4, 5)$, and $C = (-4, 1)$.
$$m_{\overline{AB}} = \frac{5-(-3)}{4-(-2)} = \frac{8}{6} = \frac{4}{3}$$
$$m_{\overline{BC}} = \frac{1-5}{-4-4} = \frac{-4}{-8} = \frac{1}{2}$$
$$m_{\overline{AC}} = \frac{1-(-3)}{-4-(-2)} = \frac{4}{-2} = -2$$

Because $m_{\overline{AC}} \cdot m_{\overline{BC}} = -1$, $\overline{AC} \perp \overline{BC}$ and $\angle C$ is

a rt. $\angle$.

19. Let $A = (3, 6)$, $B = (-6\ 4)$, and $C = (1, -2)$.

$$AB = \sqrt{(-6-3)^2 + (4-6)^2}$$
$$= \sqrt{(-9)^2 + (-2)^2} = \sqrt{81+4} = \sqrt{85}$$
$$BC = \sqrt{(1-(-6))^2 + (-2-4)^2}$$
$$= \sqrt{(-9)^2 + (-2)^2} = \sqrt{49+36} = \sqrt{85}$$
$$AC = \sqrt{(1-3)^2 + (-2-6)^2}$$
$$= \sqrt{(-2)^2 + (-8)^2} = \sqrt{4+64}$$
$$= \sqrt{68} = 2\sqrt{17}$$

Because $AB = BC$, the triangle is isosceles.

20. $R = (-5, -3)$, $S = (1, -11)$, $T = (7, -6)$, and $V = (1, 2)$.

$$m_{\overline{RS}} = \frac{-11-(-3)}{1-(-5)} = \frac{-8}{6} = \frac{-4}{3}$$
$$m_{\overline{ST}} = \frac{-6-(-11)}{7-1} = \frac{5}{6}$$
$$m_{\overline{TV}} = \frac{2-(-6)}{1-7} = \frac{8}{-6} = \frac{-4}{3}$$
$$m_{\overline{RV}} = \frac{2-(-3)}{1-(-5)} = \frac{5}{6}$$

$\therefore \overline{RS} \parallel \overline{VT}$ and $\overline{RV} \parallel \overline{ST}$ and $RSTV$ is a parallelogram.

21. Solution by graphing:
$$4x - 3y = -3$$
$$y = \frac{4}{3}x + 1$$
$$x + 2y = 13$$
$$y = -\frac{1}{2}x + \frac{13}{2}$$

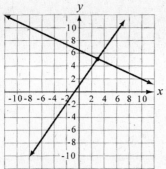

The graphs (lines) intersect at (3, 5).

22. Solution by graphing:
$$y = x + 3$$
$$y = 4x$$

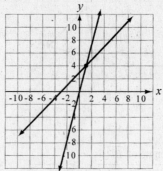

The graphs (lines) intersect at (1, 4).

23. Solution by Substitution:
$$4x - 3y = -3$$
$$x + 2y = 13$$
$$x = 13 - 2y$$
$$4(13 - 2y) - 3y = -3$$
$$52 - 8y - 3y = -3$$
$$-11y = -55$$
$$y = 5$$
$$x = 13 - 2(5) = 3$$

$(3, 5)$

24. Solution by Substitution:
$$4x = x + 3$$
$$3x = 3$$
$$x = 1$$
$$y = 4(1) = 4$$

$(1, 4)$

25.

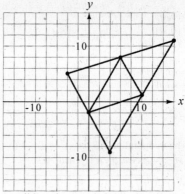

Possible points are (16, 11), (4, -9), (-4, 5).

26. a. $D = M_{\overline{AC}} = (7, 2)$

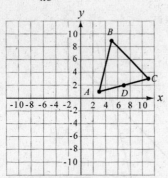

The length of $\overline{BD}$ is

$$\sqrt{(7-5)^2 + (2-9)^2} = \sqrt{2^2 + (-7)^2}$$
$$= \sqrt{4 + 49} = \sqrt{53}$$

b. $m_{\overline{AC}} = \dfrac{3-1}{11-3} = \dfrac{2}{8} = \dfrac{1}{4}$,

Then the slope of the altitude to $\overline{AC}$ is -4.

c. Since $m_{\overline{AC}} = \dfrac{1}{4}$, the slope of any line parallel

to $\overline{AC}$ is also $\dfrac{1}{4}$.

27. $A = (-a, 0)$
$B = (0, b)$
$C = (a, 0)$

28. $D = (0, 0)$
$E = (a, 0)$
$F = (a, 2a)$
$G = (0, 2a)$

29. $R = (0, 0)$
$U = (0, a)$
$T = (a, a+b)$

30. $M = (0, 0)$
$N = (a, 0)$
$Q = (a+b, c)$
$P = (b, c)$

31. a. The midpoint of $\overline{AB}$ is
$M_{\overline{AB}} = (a+c, b+d)$. Then

$$CM = \sqrt{(a+c-0)^2 + (b+d-2e)^2}$$
$$= \sqrt{(a+c)^2 + (b+d-2e)^2}$$

b. $m_{\overline{AC}} = \dfrac{2e-2b}{0-2a} = \dfrac{e-b}{-a}$ or $\dfrac{b-e}{a}$

Then the slope of the altitude to $\overline{AC}$ is

$-\dfrac{a}{b-e}$ or $\dfrac{b-e}{a}$.

c. The altitude from B to $\overline{AC}$ contains the point

$(2c, 2d)$ and has slope $m = \dfrac{a}{e-b}$ from part

(b). $y - 2d = \dfrac{a}{e-b}(x - 2c)$.

32. See Section 10.4, #18.

33. If the diagonals of a rectangle are perpendicular, then the rectangle is a square.

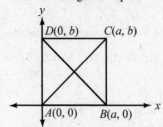

Proof: Let rect. *ABCD* have vertices as shown.
If $\overline{DB} \perp \overline{AC}$, then $m_{\overline{DB}} \cdot m_{\overline{AC}} = -1$. But

$m_{\overline{DB}} = \dfrac{0-b}{a-0} = \dfrac{-b}{a}$ and $m_{\overline{AC}} = \dfrac{b-0}{a-0} = \dfrac{b}{a}$. Then

$$\dfrac{-b}{a} \cdot \dfrac{b}{a} = -1$$
$$\dfrac{-b^2}{a^2} = -1$$
$$-b^2 = -a^2$$
$$b^2 = a^2$$

Since a and b are both positive, $a = b$. Then
$AB = a - 0 = a$ and $AD = b - 0 = b$. Since $a = b$,
$AB = AD$. If $AB = AD$, then *ABCD* is a square.

34. If the diagonals of a trapezoid are equal in length, then the trapezoid is an isosceles trapezoid.

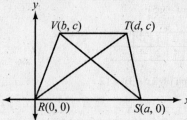

Proof: Let trap. *RSTV* have vertices as shown. If $RT = VS$, then

$$\sqrt{(d-0)^2 + (c-0)^2} = \sqrt{(a-b)^2 + (0-c)^2}$$
$$\sqrt{d^2 + c^2} = \sqrt{(a-b)^2 + (-c)^2}$$
$$\sqrt{d^2 + c^2} = \sqrt{(a-b)^2 + c^2}$$

Squaring, $d^2 + c^2 = (a-b)^2 + c^2$
$$d^2 = (a-b)^2$$
$$d = (a-b$$

Comparing the lengths of $\overline{RV}$ and $\overline{ST}$,

$$RV = \sqrt{(b-0)^2 + (c-0)^2} = \sqrt{b^2 + c^2}$$
$$ST = \sqrt{(a-d)^2 + (0-c)^2} = \sqrt{(a-d)^2 + c^2}$$

Now *RV* would equal *ST* if
$$b^2 + c^2 = (a-d)^2 + c^2$$
$$b^2 = (a-d)^2$$
$$b = a-d$$

Because $d = a-b$ leads to $b = a-d$, so $RV = ST$. Then *RSTV* is isosceles.

35. If two medians of a triangle are equal in length, then the triangle is isosceles.

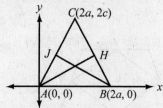

Proof: Let $\triangle ABC$ has vertices as shown, so that the midpoint of $\overline{AC}$ is $J = (b, c)$ and the midpoint of $\overline{BC}$ is $H = (a+b, c)$. If $AH = BJ$, then

$$\sqrt{(a+b-0)^2 + (c-0)^2} = \sqrt{(2a-b)^2 + (0-c)^2}$$
$$\sqrt{(a+b)^2 + c^2} = \sqrt{(2a-b)^2 + c^2}$$

Squaring, $(a+b)^2 + c^2 = (2a-b)^2 + c^2$
$$(a+b)^2 = (2a-b)^2$$

Taking the principal of square roots,
$$a+b = 2a-b$$
$$2b = a$$

Then point $C = (a, 2c)$.

Now $AC = \sqrt{(a-0)^2 + (2c-0)^2}$
$$= \sqrt{a^2 + (2c)^2} = \sqrt{a^2 + 4c^2}$$

and $BC = \sqrt{(2a-a)^2 + (0-2c)^2}$
$$= \sqrt{a^2 + (-2c)^2} = \sqrt{a^2 + 4c^2}$$

Because $AC = BC$, the triangle is isosceles.

36. The segments joining the midpoints of consecutive sides of an isosceles trapezoid form a rhombus.

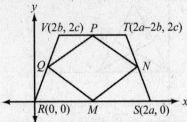

Proof: Let trapezoid *RSTV* have vertices as shown so that $RV = TS$. Where *M*, *N*, *P*, and *Q* are the midpoints of the sides,

$$M = \left(\frac{0+2a}{2}, \frac{0+0}{2}\right) = (a, 0)$$

$$N = \left(\frac{(2a-2b)+2a}{2}, \frac{0+2c}{2}\right) = (2a-b, c)$$

$$P = \left(\frac{(2a-2b)+2b}{2}, \frac{2c+2c}{2}\right) = (a, 2c)$$

$$Q = \left(\frac{0+2b}{2}, \frac{0+2c}{2}\right) = (b, c)$$

By an earlier theorem, *MNPQ* is a parallelogram. We must show that two adjacent sides of parallelogram *MNPQ* are equal in length.

$$QM = \sqrt{(a-b)^2 + (0-c)^2}$$
$$= \sqrt{(a-b)^2 + (-c)^2} = \sqrt{(a-b)^2 + c^2}$$
$$MN = \sqrt{(2a-b-a)^2 + (c-0)^2} = \sqrt{(a-b)^2 + c^2}$$

Now $QM = MN$, so *MNPQ* is a rhombus.

CHAPTER TEST

1. a. (5, -3)

   b. (0, -4)

2.

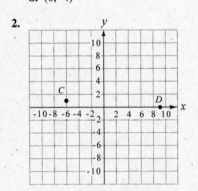

3. $d = \sqrt{(-6-0)^2 + (1-9)^2} = \sqrt{(-6)^2 + (-8)^2}$
$$= \sqrt{36+64} = \sqrt{100} = 10$$

4. $M = \left(\dfrac{x_1 + x_2}{2}, \dfrac{y_1 + y_2}{2}\right) = \left(\dfrac{-6+0}{2}, \dfrac{1+9}{2}\right)$
$$= (-3, 5)$$

5.

x	0	3	0	9
y	4	2	4	-2

6. Graph of $2x + 3y = 12$,

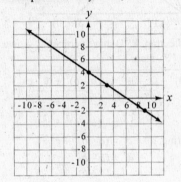

7. a. $m = \dfrac{-6-3}{2-(-1)} = \dfrac{-9}{3} = -3$

   b. $m = \dfrac{d-b}{c-a}$

8. a. $\dfrac{2}{3}$

   b. $-\dfrac{3}{2}$

9. $AB = \sqrt{(0-0)^2 + (a-0)^2} = \sqrt{a^2} = a$
$BC = \sqrt{(a+b-a)^2 + (c-0)^2} = \sqrt{b^2 + c^2}$
$CD = \sqrt{(b-(a+b))^2 + (c-c)^2} = \sqrt{(-a)^2} = a$
$AD = \sqrt{(b-0)^2 + (c-0)^2} = \sqrt{b^2 + c^2}$

Since $\overline{AB} = \overline{CD}$ and $\overline{BC} = \overline{AD}$, *ABCD* is a parallelogram.

10. $AB = \sqrt{(0-0)^2 + (a-0)^2} = \sqrt{a^2} = a$
$AD = \sqrt{(b-0)^2 + (c-0)^2} = \sqrt{b^2 + c^2}$

If $\overline{AB} = \overline{AD}$, then $a = \sqrt{b^2 + c^2}$ or $a^2 = b^2 + c^2$.

11. a. isosceles triangle

   b. trapezoid

12. a. Slope Formula

   b. Distance Formula

13. Let $D = (0, 0)$ and $E = (2a, 0)$, then $F = (a, b)$.

14. (b)

15. Since *RSTV* is a parallelogram, then $m_{\overline{VR}} = m_{\overline{TS}}$.

$$m_{\overline{VR}} = \frac{v-0}{0-r} = -\frac{v}{r}$$

$$m_{\overline{TS}} = \frac{v-0}{t-s} = \frac{v}{t-s}$$

So $-\frac{v}{r} = \frac{v}{t-s}$

$$rv = -v(t-s)$$

$$r = -t+s$$

or $s = r+t$ or $t = -r+s$.

16. a. $m = \frac{6-4}{2-0} = \frac{2}{2} = 1$

$$y-4 = 1(x-0)$$

$$y-4 = x$$

$$y = x+4$$

b. $m = \frac{3}{4}$, $b = -3$

$$y = \frac{3}{4}x - 3$$

17. The slope of the perpendicular line $m = c$.

$$y - b = c(x-a)$$

$$y - b = cx - ac$$

$$y = cx + (b-ac)$$

18. $x + 2y = 6$

$$x = -2y + 6$$

Use substitution,

$$2x - y = 7$$

$$2(-2y+6) - y = 7$$

$$-4y + 12 - y = 7$$

$$-5y = -5$$

$$y = 1$$

$$x = -2(1) + 6 = 4$$

$(4, 1)$

19. $5x - 2y = -13$ Multiply by 5

     $3x + 5y = 17$ Multiply by 2

     $25x - 10y = -65$

     $\underline{6x + 10y = 34}$

          $31x = -31$

             $x = -1$

     $3(-1) + 5y = 17$

       $-3 + 5y = 17$

          $5y = 20$

           $y = 4$

$(-1, 4)$

20. $M = (a+b, c)$ and $N = (a, 0)$.

Then $m_{\overline{AC}} = \frac{2c-0}{2b-0} = \frac{c}{b}$.

Also $m_{\overline{MN}} = \frac{c-0}{a+b-a} = \frac{c}{b}$.

With $m_{\overline{AC}} = m_{\overline{MN}}$, it follows that $\overline{AC} \parallel \overline{MN}$.

Chapter 11: Introduction to Trigonometry

SECTION 11.1: The Sine Ratio and Applications

1. $\sin\alpha = \dfrac{\text{opposite}}{\text{hypotenuse}} = \dfrac{5}{13}$

$\sin\beta = \dfrac{12}{13}$

5.
$$a^2 + b^2 = c^2$$
$$\left(\sqrt{2}\right)^2 + \left(\sqrt{3}\right)^2 = c^2$$
$$2 + 3 = c^2$$
$$5 = c^2$$
$$\sqrt{5} = c$$

$\sin\alpha = \dfrac{\sqrt{3}}{\sqrt{5}} = \dfrac{\sqrt{3}}{\sqrt{5}} \cdot \dfrac{\sqrt{5}}{\sqrt{5}} = \dfrac{\sqrt{15}}{5}$

$\sin\beta = \dfrac{\sqrt{2}}{\sqrt{5}} = \dfrac{\sqrt{2}}{\sqrt{5}} \cdot \dfrac{\sqrt{5}}{\sqrt{5}} = \dfrac{\sqrt{10}}{5}$

9. $\sin 17° = 0.2924$

13. $\sin 72° = 0.9511$

17. $\sin 43° = \dfrac{a}{16}$

$a = 16\sin 43°$

$a \approx 16(0.6820)$

$a \approx 10.9 \text{ ft}$

$\sin 47° = \dfrac{b}{16}$

$b = 16\sin 47°$

$b \approx 16(0.7314)$

$b \approx 11.7 \text{ ft}$

21. $\sin\alpha = \dfrac{12}{25} = 0.4800$

$\alpha \approx 29°$

$\beta \approx 90° - 29°$ or $\beta \approx 61°$

25. $\sin\alpha = \dfrac{x}{3x} = \dfrac{1}{3} \approx 0.3333$

$\alpha \approx 19°$

$\beta \approx 90° - 19°$ or $\beta \approx 71°$

29. Let d represent the distance between Danny and the balloon.

$\sin 75° = \dfrac{100}{d}$

$d \cdot \sin 75° = 100$

$d = \dfrac{100}{\sin 75°}$

$d \approx \dfrac{100}{0.9659}$

$d \approx 103.5 \text{ ft}$

33. $\sin\alpha = \dfrac{4}{10}$

$\sin\alpha = 0.4000$

$\alpha \approx 24°$

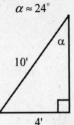

37. $\sin\theta = \dfrac{10}{13} \approx 0.7692$

$\theta \approx 50°$

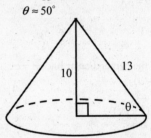

SECTION 11.2: The Cosine Ratio and Applications

1. $\cos\alpha = \dfrac{\text{adjacent}}{\text{hypotenuse}} = \dfrac{12}{13}$

$\cos\beta = \dfrac{5}{13}$

5.
$$a^2 + b^2 = c^2$$
$$\left(\sqrt{3}\right)^2 + \left(\sqrt{2}\right)^2 = c^2$$
$$3 + 2 = c^2$$
$$5 = c^2$$
$$\sqrt{5} = c$$

$\cos\alpha = \dfrac{\sqrt{2}}{\sqrt{5}} = \dfrac{\sqrt{2}}{\sqrt{5}} \cdot \dfrac{\sqrt{5}}{\sqrt{5}} = \dfrac{\sqrt{10}}{5}$

$\cos\beta = \dfrac{\sqrt{3}}{\sqrt{5}} = \dfrac{\sqrt{3}}{\sqrt{5}} \cdot \dfrac{\sqrt{5}}{\sqrt{5}} = \dfrac{\sqrt{15}}{5}$

9. $\cos 23° \approx 0.9205$

13. $\cos 90° = 0$

17. $\cos 32° = \dfrac{a}{100}$

$\quad a = 100 \cos 32°$

$\quad a \approx 100(0.8480)$

$\quad a \approx 84.8$ ft

$\sin 32° = \dfrac{b}{100}$

$\quad b = 100 \sin 32°$

$\quad b \approx 100(0.5299)$

$\quad b \approx 53.0$ ft

21. $\cos 51° = \dfrac{12}{c}$

$\quad c = \dfrac{12}{\cos 51°}$

$\quad c \approx \dfrac{12}{0.6293}$

$\quad c \approx 19.1$ ft

$\sin 51° = \dfrac{d}{c}$

$\quad d = c \cdot \sin 51°$

$\quad d \approx (19.1)(0.7771)$

$\quad d \approx 14.8$ ft

25. $a^2 + b^2 = c^2$

$\quad \left(\sqrt{3}\right)^2 + \left(\sqrt{2}\right)^2 = c^2$

$\quad 3 + 2 = c^2$

$\quad 5 = c^2$

$\quad \sqrt{5} = c$

$\cos \alpha = \dfrac{\sqrt{2}}{\sqrt{5}} = \sqrt{\dfrac{2}{5}} = \sqrt{0.4} \approx 0.6325$

$\quad \alpha \approx 51°$

$\cos \beta \approx 90° - 51°$

$\quad \beta \approx 39°$

29. $\cos \theta = \dfrac{10}{12} \approx 0.8333$

$\quad \theta \approx 34°$

33. Let c represent the measure of the central angle of the regular pentagon. Then $c = \dfrac{360°}{5} = 72°$.

Because the apothem shown bisects the central angle,

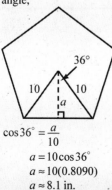

$\cos 36° = \dfrac{a}{10}$

$\quad a = 10 \cos 36°$

$\quad a \approx 10(0.8090)$

$\quad a \approx 8.1$ in.

37. Let d represent the length of the diagonal of the base and x the length of an edge of the cube. By the Pythagorean Theorem,

$\quad d^2 = x^2 + x^2$

$\quad d^2 = 2x^2.$

Let D represent the Pythagorean Theorem again,

$\quad x^2 + d^2 = D^2$

$\quad x^2 + 2x^2 = D^2$

so that

$\quad D^2 = 3x^2$

$\quad D = \sqrt{3x^2}$

$\quad D = x\sqrt{3}$

Now, $\cos \alpha = \dfrac{\text{adjacent}}{\text{hypotenuse}} = \dfrac{x}{x\sqrt{3}} = \dfrac{1}{\sqrt{3}}$

$\quad \cos \alpha \approx 0.5774$

$\quad \alpha \approx 55°$

41.

$\sin \theta = \dfrac{\frac{1}{2}b}{s}$

$\quad s \cdot \sin \theta = \dfrac{1}{2}b$

$\quad b = 2s \cdot \sin \theta$

$\cos \theta = \dfrac{h}{s}$

$\quad h = s \cdot \cos \theta$

$A_\triangle = \dfrac{1}{2}bh$

$A_\triangle = \dfrac{1}{2}(2s \cdot \sin \theta)(s \cdot \cos \theta)$

$A_\triangle = s^2 \cdot \sin \theta \cdot \cos \theta$

SECTION 11.3: The Tangent Ratio and Other Ratios

1. $\tan \alpha = \dfrac{\text{opposite}}{\text{adjacent}} = \dfrac{3}{4}$

$\quad \tan \beta = \dfrac{4}{3}$

5. Using the Pythagorean Triple, (5, 12, 13), $b = 12$.

$$\sin \alpha = \frac{\text{opposite}}{\text{hypotenuse}} = \frac{5}{13}$$

$$\cos \alpha = \frac{\text{adjacent}}{\text{hypotenuse}} = \frac{12}{13}$$

$$\tan \alpha = \frac{\text{opposite}}{\text{adjacent}} = \frac{5}{12}$$

$$\cot \alpha = \frac{\text{adjacent}}{\text{opposite}} = \frac{12}{5}$$

$$\sec \alpha = \frac{\text{hypotenuse}}{\text{adjacent}} = \frac{13}{12}$$

$$\csc \alpha = \frac{\text{hypotenuse}}{\text{opposite}} = \frac{13}{5}$$

9. $a^2 + b^2 = c^2$

$$x^2 + b^2 = \left(\sqrt{x^2 + 1}\right)^2$$
$$x^2 + b^2 = x^2 + 1$$
$$b^2 = 1$$
$$b = 1$$

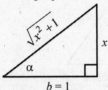

$$\sin \alpha = \frac{x}{\sqrt{x^2 + 1}}$$

$$= \frac{x}{\sqrt{x^2 + 1}} \cdot \frac{\sqrt{x^2 + 1}}{\sqrt{x^2 + 1}}$$

$$= \frac{x\sqrt{x^2 + 1}}{x^2 + 1}$$

$$\cos \alpha = \frac{1}{\sqrt{x^2 + 1}}$$

$$= \frac{1}{\sqrt{x^2 + 1}} \cdot \frac{\sqrt{x^2 + 1}}{\sqrt{x^2 + 1}}$$

$$= \frac{\sqrt{x^2 + 1}}{x^2 + 1}$$

$$\tan \alpha = \frac{x}{1}$$

$$\cot \alpha = \frac{1}{x}$$

$$\sec \alpha = \frac{\sqrt{x^2 + 1}}{1} = \sqrt{x^2 + 1}$$

$$\csc \alpha = \frac{\sqrt{x^2 + 1}}{x}$$

13. $\tan 57° = 1.5399$

17. $\cos 58° = \dfrac{y}{10}$

$$y = 10 \cos 58°$$
$$y \approx 10(0.5299)$$
$$y \approx 5.3$$

$$\sin 58° = \frac{z}{10}$$
$$z = 10 \sin 58°$$
$$z \approx 10(0.8480)$$
$$z \approx 8.5$$

21. $\tan \alpha = \dfrac{3}{4} = 0.7500$

$$\alpha \approx 37°$$
$$\beta \approx 90° - 37° \text{ or } \beta \approx 53°$$

25. $\tan \alpha = \dfrac{\sqrt{5}}{4} \approx \dfrac{2.2361}{4} \approx 0.5590$

$$\alpha \approx 29°$$
$$\beta \approx 90° - 29° \text{ or } \beta \approx 61°$$

29. $\csc 30° = \dfrac{1}{\sin 30°} = \dfrac{1}{0.5} = 2.0000$ (exact)

33. a. $\sin \alpha = \dfrac{a}{c}$ and $\cos \alpha = \dfrac{b}{c}$

Then $\dfrac{\sin \alpha}{\cos \alpha} = \sin \alpha \div \cos \alpha = \dfrac{a}{c} \div \dfrac{b}{c}$.

In turn, $\dfrac{\sin \alpha}{\cos \alpha} = \dfrac{a}{c} \cdot \dfrac{c}{b} = \dfrac{a}{b} = \tan \alpha$.

b. $\tan 23° \approx 0.4245$

$$\frac{\sin 23°}{\cos 23°} \approx \frac{0.39073}{0.92050} \approx 0.4245 \approx \tan 23°$$

37. $\sin 5° = \dfrac{120}{x}$

$$x \cdot \sin 5° = 120$$
$$x = \frac{120}{\sin 5°}$$
$$x \approx \frac{120}{0.0872} \approx 1376.8 \text{ ft}$$

41

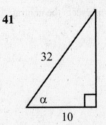

$$\cos \alpha = \frac{10}{32}$$
$$\cos \alpha = 0.3125$$
$$\alpha \approx 72°$$

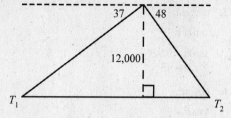

45   37   48

12,000

T_1 T_2

$$\tan 37 = \frac{12,000}{x}$$
$$x \cdot \tan 37 = 12,000$$
$$x = \frac{12,000}{\tan 37}$$
$$x \approx 15,924.5$$

$$\tan 48 = \frac{12,000}{y}$$
$$y \cdot \tan 48 = 12,000$$
$$y = \frac{12,000}{\tan 48}$$
$$y \approx 10,804.8$$

$$\text{Distance} \approx 15,924.5 + 10,804.8$$
$$\approx 26,729.3 \approx 26,730 \text{ feet}$$

SECTION 11.4: Applications with Acute Triangles

1. a. $A = \frac{1}{2}ab\sin\gamma$

$A = \frac{1}{2} \cdot 5 \cdot 6 \cdot \sin 78°$

b. $\alpha + \beta + \gamma = 180°$
$36° + 88° + \gamma = 180°$
$124 + \gamma = 180$
$\gamma = 56°$

$A = \frac{1}{2}ab\sin\gamma$

$A = \frac{1}{2} \cdot 5 \cdot 7 \cdot \sin 56°$

5. a. $c^2 = a^2 + b^2 - 2ab\cos\gamma$
$c^2 = (5.2)^2 + (7.9)^2 - 2(5.2)(7.9)\cos 83°$

b. $a^2 = b^2 + c^2 - 2bc\cos\alpha$
$6^2 = 9^2 + 10^2 - 2(9)(10)\cos\alpha$

9. a. (3, 4, 5) is a Pythagorean Triple. γ lies opposite the longest side and must be a right angle.

b. 90°

13. With measures of angles shown, the third angle measures 70°. Then the triangle is isosceles, with sides as shown.

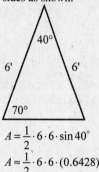

40°

6'   6'

70°

$A = \frac{1}{2} \cdot 6 \cdot 6 \cdot \sin 40°$

$A \approx \frac{1}{2} \cdot 6 \cdot 6 \cdot (0.6428)$

$A \approx 11.6 \text{ ft}^2$

17. $\frac{\sin\alpha}{a} = \frac{\sin\beta}{b}$

$\frac{\sin 60°}{x} = \frac{\sin 70°}{12}$

$\frac{0.8660}{x} = \frac{0.9397}{12}$

$x(0.9397) = 12(0.8660)$

$x \approx 11.1 \text{ in.}$

21. $\frac{\sin\gamma}{10} = \frac{\sin 80°}{12}$

$\frac{\sin\gamma}{10} = \frac{0.9397}{12}$

$12(\sin\gamma) = 10(0.9848)$

$\sin\gamma \approx 0.8207$

$\gamma \approx 55°$

25. $x^2 = 8^2 + 12^2 - 2 \cdot 8 \cdot 12\cos 60°$

$x^2 = 64 + 144 - 192\cos 60°$

$x^2 = 208 - 192(0.5)$

$x^2 = 112$

$x = \sqrt{112} \approx 10.6$

29. a. $x^2 = 150^2 + 180^2 - 2(150)(180)\cos 80°$

$x^2 = 22,500 + 32,400 - 54,000(0.1736)$

$x^2 = 44,525.6$

$x = \sqrt{44,525.6}$

$x \approx 213.4 \text{ feet}$

150'   x

80°

180'

b. $A = \frac{1}{2}(150)(180)\sin 80°$

$A \approx \frac{1}{2}(150)(180)(0.9848)$

$A \approx 13,294.9 \text{ ft}^2$

33. The third angle measures 85°.
Using the Law of Sines,

$$\frac{\sin 85°}{x} = \frac{\sin 30°}{8}$$

$$\frac{0.9962}{x} = \frac{0.5}{8}$$

$$0.5x = 8(0.9962)$$

$$x \approx 15.9 \text{ ft}$$

37. $a^2 = 27^2 + 27^2 - 2(27)(27)\cos 30°$

$$a^2 = 729 + 729 - 2(729) \cdot \frac{\sqrt{3}}{2}$$

$$a^2 \approx 195.33$$

$$a \approx 13.97 \approx 14.0 \text{ feet}$$

41. $A = (6.3)(8.9)\sin 67.5°$

$$A = 51.8 \text{ cm}^2$$

CHAPTER REVIEW

1. sine;

$$\sin 40° = \frac{a}{16}$$

$$a = 16\sin 40°$$

$$a \approx 10.3 \text{ in.}$$

2. sine;

$$\sin 70° = \frac{d}{8}$$

$$d = 8\sin 70°$$

$$d \approx 7.5 \text{ ft}$$

3. cosine;

$$\cos 80° = \frac{4}{c}$$

$$c = \frac{4}{\cos 80°}$$

$$c \approx \frac{4}{0.1736}$$

$$c \approx 23 \text{ in.}$$

4. sine;

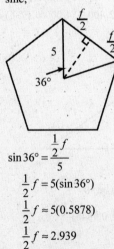

$$\sin 36° = \frac{\frac{1}{2}f}{5}$$

$$\frac{1}{2}f = 5(\sin 36°)$$

$$\frac{1}{2}f \approx 5(0.5878)$$

$$\frac{1}{2}f \approx 2.939$$

$$f \approx 5.9 \text{ ft}$$

5. tangent;

$$\tan\alpha = \frac{13}{14}$$

$$\tan\alpha \approx 0.9286$$

$$\alpha \approx 43°$$

6. cosine;

$$\cos\theta = \frac{8}{15}$$

$$\cos\theta = 0.5333$$

$$\theta \approx 58°$$

7. sine;

$$\sin\alpha = \frac{9}{12}$$

$$\sin\alpha \approx 0.7500$$

$$\alpha \approx 49°$$

8. tangent;

$$\tan\beta = \frac{7}{24}$$

$$\tan\beta \approx 0.2917$$

$$\beta \approx 16°$$

9. Law of Sines

$$\frac{\sin 57°}{x} = \frac{\sin 49°}{8}$$

$$x(\sin 49°) = 8(\sin 57°)$$

$$x(0.7547) = 8(0.8387)$$

$$x \approx 8.9 \text{ units}$$

10. Law of Cosines

$$15^2 = 14^2 + 16^2 - 2(14)(16)\cos\alpha$$

$$225 = 196 + 256 - 448\cos\alpha$$

$$448\cos\alpha = 227$$

$$\cos\alpha = \frac{227}{448}$$

$$\cos\alpha \approx 0.5067$$

$$\alpha \approx 60°$$

11. The third angle measures 80°.
Law of Sines

$$\frac{\sin 40°}{y} = \frac{\sin 80°}{20}$$

$$y(\sin 80°) = 20(\sin 40°)$$

$$y(0.9848) = 20(0.6428)$$

$$y \approx 13.1$$

12. Law of Cosines

$$w^2 = 14^2 + 21^2 - 2(14)(21)\cos 60°$$

$$w^2 = 196 + 441 - 294$$

$$w^2 = 343$$

$$w = \sqrt{343}$$

$$w \approx 18.5$$

13. The remaining angles of the triangle measure 47° and 74°.

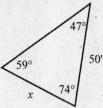

Using the Law of Sines,

$$\frac{\sin 47°}{x} = \frac{\sin 59°}{50}$$

$$\frac{0.7314}{x} = \frac{0.8572}{50}$$

$$x(0.8572) = 50(0.7314)$$

$$x \approx 42.7 \text{ feet}$$

14. Let d be the length of the shorter diagonal.

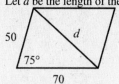

Law of Cosines

$$d^2 = 50^2 + 70^2 - 2(50)(70)\cos 75°$$

$$d^2 = 2500 + 4900 - 7000(0.2588)$$

$$d^2 = 2500 + 4900 - 1811.6$$

$$d^2 = 5588.4$$

$$d = \sqrt{5588.4}$$

$$d \approx 74.8 \text{ cm}$$

15. Law of Cosines

$$6^2 = 6^2 + 11^2 - 2(6)(11)\cos\alpha$$

$$36 = 366 + 121 - 132\cos\alpha$$

$$132\cos\alpha = 121$$

$$\cos\alpha = \frac{121}{132}$$

$$\cos\alpha \approx 0.9167$$

$$\alpha \approx 23.6°$$

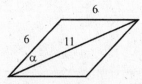

The acute angle of the rhombus measures $2\alpha \approx 47°$.

16.

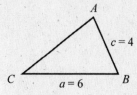

$$A = \frac{1}{2}ac\sin B$$

$$9.7 = \frac{1}{2}(6)(4)\sin B$$

$$9.7 = 12\sin B$$

$$\sin B = \frac{9.7}{12} \approx 0.8083$$

$$B \approx 54°$$

17.

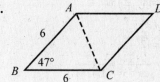

If the acute angle measures 47°, then

$A = \frac{1}{2}ab\sin 47°$ gives one-half the desired area.

$$A = \frac{1}{2} \cdot 6 \cdot 6 \cdot \sin 47°$$

$$A \approx 13.16 \text{ in}^2$$

for $\triangle ABC$. The area of rhombus $ABCD$ is approximately 26.3 in^2.

18. If $m\angle R = 45°$,

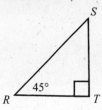

then $m\angle S = 45°$ also. Then $\overline{RT} = \overline{ST}$. Let

$RT = ST = x$. Now, $\tan R = \tan 45° = \frac{x}{x} = 1$.

19. If $m\angle S = 30°$,

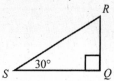

then the sides of $\triangle RQS$ can be represented by

$RQ = x$, $RS = 2x$, and $SQ = x \cdot \sqrt{3}$.

$$\sin S = \sin 30° = \frac{x}{2x} = \frac{1}{2}.$$

20. If $m\angle T = 60°$,

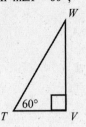

then the sides of $\triangle TVW$ can be represented by
$TV = x$, $TW = 2x$, and $VW = x\sqrt{3}$.

$\sin T = \sin 60° = \dfrac{x\sqrt{3}}{2x} = \dfrac{\sqrt{3}}{2}$.

21. Because alt. int. $\angle$s are $\cong$,

$$\tan 55° = \dfrac{12}{x}$$
$$x(\tan 55°) = 12$$
$$x = \dfrac{12}{\tan 55°} \approx \dfrac{12}{1.4281}$$
$$x \approx 8.4 \text{ ft}$$

22. $\sin 60° = \dfrac{x}{200}$

$$x = 200 \sin 60°$$
$$x \approx 200(0.866)$$
$$x \approx 173.2$$

If the rocket rises 173.2 feet per second, then its altitude after 5 seconds will be approximately 866 feet.

23. $\cos \alpha = \dfrac{3}{4}$

$$\cos \alpha = 0.75$$
$$\alpha \approx 41°$$

24. $\sin \alpha = \dfrac{300}{2200}$

$$\sin \alpha = 0.1364$$
$$\alpha \approx 8°$$

25. Let x represent one-half the length of a side of a regular pentagon.

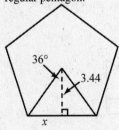

$$\tan 36° = \dfrac{x}{3.44}$$
$$x = 3.44(\tan 36°)$$
$$x \approx 3.44(0.7265)$$
$$x \approx 2.50$$

Then the length of each side is approximately 5.0 cm.

26.

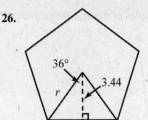

$$\cos 36° = \dfrac{3.44}{r}$$
$$r = \dfrac{3.44}{\cos 36°}$$
$$r \approx \dfrac{3.44}{0.8090}$$
$$r \approx 4.3 \text{ cm}$$

27. The altitude bisects the base where β represents the measure of the base angle,

$$\cos \beta = \dfrac{15}{40}$$
$$\cos \beta = 0.375$$
$$\beta \approx 68°$$

28. The measure of acute angle α is one-half the desired angle's measure.

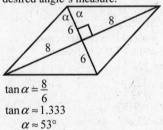

$$\tan \alpha = \dfrac{8}{6}$$
$$\tan \alpha \approx 1.333$$
$$\alpha \approx 53°$$

29.
$$\tan 23° = \dfrac{a}{b}$$
$$\tan 23° = \dfrac{3}{b}$$
$$b(\tan 23°) = 3$$
$$b = \dfrac{3}{\tan 23°}$$
$$b \approx \dfrac{3}{0.4245} \approx 7$$

The grade of the hill is 3 to 7 (or 3:7).

30. Let S_1 represent the distance to the nearer ship and S_2 represent the distance to the farther ship.

$$\tan 32° = \frac{2500}{S_2} \quad \text{and} \quad \tan 44° = \frac{2500}{S_1}$$

$$S_2 = \frac{2500}{\tan 32°} \qquad S_1 = \frac{2500}{\tan 44°}$$

$$S_2 \approx 4000.8 \qquad S_1 \approx 2588.8$$

The distance between the ships is 4000.8 − 2588.8 or approximately 1412.0 meters.

31. $\sin \theta = \dfrac{7}{25}$

Using $\sin^2 \theta + \cos^2 \theta = 1$,

$$\left(\frac{7}{25}\right)^2 + \cos^2 \theta = 1$$

$$\frac{49}{625} + \cos^2 \theta = 1$$

$$\cos^2 \theta = 1 - \frac{49}{625}$$

$$\cos^2 \theta = \frac{576}{625}$$

$$\cos \theta = \sqrt{\frac{576}{625}} = \frac{\sqrt{576}}{\sqrt{625}} = \frac{24}{25}$$

Because $\sec \theta = \dfrac{1}{\cos \theta}$,

$$\sec \theta = \frac{1}{\frac{24}{25}} = \frac{25}{24}.$$

32. $\tan \theta = \dfrac{11}{60}$

Using $\tan^2 \theta + 1 = \sec^2 \theta$,

$$\left(\frac{11}{60}\right)^2 + 1 = \sec^2 \theta$$

$$\frac{121}{3600} + 1 = \sec^2 \theta$$

$$\frac{3721}{3600} = \sec^2 \theta$$

$$\sec \theta = \sqrt{\frac{3721}{3600}} = \frac{\sqrt{3721}}{\sqrt{3600}} = \frac{61}{60}$$

Because $\cot \theta = \dfrac{1}{\tan \theta}$,

$$\cot \theta = \frac{1}{\frac{11}{60}} = \frac{60}{11}.$$

33. $\cot \theta = \dfrac{21}{20}$

Using $\cot^2 \theta + 1 = \csc^2 \theta$,

$$\left(\frac{21}{20}\right)^2 + 1 = \csc^2 \theta$$

$$\frac{841}{400} = \csc^2 \theta$$

$$\csc \theta = \sqrt{\frac{841}{400}} = \frac{\sqrt{841}}{\sqrt{400}} = \frac{29}{20}$$

Because $\sin \theta = \dfrac{1}{\csc \theta}$,

$$\sin \theta = \frac{1}{\frac{29}{20}} = \frac{20}{29}.$$

34.

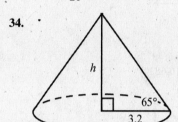

$$\tan 65° = \frac{h}{3.2}$$

$$h = 3.2 \cdot \tan 65$$

$$h \approx 6.9 \text{ feet}$$

$$V = \frac{1}{3} B h$$

$$V = \frac{1}{3} \pi r^2 h$$

$$V \approx \frac{1}{3} \pi \cdot (3.2)^2 \cdot (6.9)$$

$$V \approx 74.0 \text{ ft}^3$$

CHAPTER TEST

1. a. $\sin \alpha = \dfrac{a}{c}$

   b. $\tan \beta = \dfrac{b}{a}$

2. a. $\cos \beta = \dfrac{6}{10} = \dfrac{3}{5}$

   b. $\sin \alpha = \dfrac{6}{10} = \dfrac{3}{5}$

3. a. $\tan 45° = \dfrac{1}{1} = 1$

   b. $\tan 60° = \dfrac{\sqrt{3}}{2}$

4. a. $\sin 23° \approx 0.3907$

b. $\cos 79° \approx 0.1908$

5. $\sin \theta = 0.6691$
$\quad\quad \theta \approx 42°$

6. a. $\tan 26°$

b. $\cos 47°$

7.

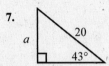

$\sin 43° = \dfrac{a}{20}$
$\quad a = 20 \sin 43°$
$\quad a \approx 20(0.6820)$
$\quad a \approx 14$

8.

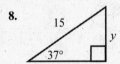

$\sin 37° = \dfrac{y}{15}$
$\quad y = 15 \sin 37°$
$\quad y \approx 15(0.6018)$
$\quad y \approx 9$

9.

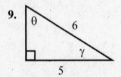

$\sin \theta = \dfrac{5}{6}$
$\sin \theta \approx 0.8333$
$\quad \theta \approx 56°$

10. a. $\cos \beta = \dfrac{a}{c} = \sin \alpha$
True

b. True

11. $\sin 67° = \dfrac{x}{100}$
$\quad x = 100 \sin 67°$
$\quad x \approx 100(0.9205)$
$\quad x \approx 92$ feet

12. $\sin \theta = \dfrac{2}{12}$
$\sin \theta \approx 0.1667$
$\quad \theta \approx 10°$

13. a. $\csc \alpha = \dfrac{1}{\sin \alpha} = \dfrac{1}{\frac{1}{2}} = 2$

b. $\sin \alpha = \dfrac{1}{2}$
$\sin \alpha = 0.5$
$\quad \alpha = 30°$

14. a.

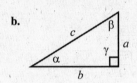

$\cos \beta = \dfrac{a}{c}$
$\sin \alpha = \dfrac{a}{c}$

b.

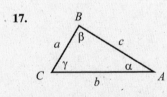

$\cos \beta = \dfrac{a}{c}$
$\sec \beta = \dfrac{1}{\cos \beta}$
$\sec \beta = \dfrac{1}{\frac{a}{c}} = \dfrac{c}{a}$

15. $A = \dfrac{1}{2} bc \sin \alpha$
$A = \dfrac{1}{2}(8)(12) \sin 60°$
$A \approx 48(0.8660)$
$A \approx 42 \text{ cm}^2$

16.

$\dfrac{\sin \alpha}{a} = \dfrac{\sin \beta}{b} = \dfrac{\sin \gamma}{c}$

17.

$a^2 = b^2 + c^2 - 2bc \cos \alpha$

18. Law of Sines
$$\frac{\sin \alpha}{a} = \frac{\sin \beta}{b}$$
$$\frac{\sin 65^\circ}{10} = \frac{\sin \alpha}{6}$$
$$10 \sin \alpha = 6 \sin 65$$
$$\sin \alpha = \frac{6 \sin 65}{10}$$
$$\sin \alpha \approx 0.5438$$
$$\alpha \approx 33^\circ$$

19. Law of Cosines
$$x^2 = 12^2 + 8^2 - 2(12)(8) \cos 60^\circ$$
$$x^2 = 144 + 64 - 192(0.5)$$
$$x^2 = 112$$
$$x = \sqrt{112}$$
$$x \approx 11$$

20.

In an isosceles triangle formed by 2 radii
and a side of the pentagon, draw the
height of the triangle to the side. This is the
apothem with length, a. If *s* represents
the side length of the pentagon (the base of
the isosceles triangle), then

$$\tan 54 = \frac{a}{\frac{1}{2}s}$$

$$\frac{1}{2}s \cdot \tan 54 = a$$

$$s = \frac{2a}{\tan 54}$$

$$A_{1\Delta} = \frac{1}{2} \cdot \frac{2a}{\tan 54} \cdot a$$

$$A_{1\Delta} = \frac{a^2}{\tan 54}$$

$$A_{ABCDE} = \frac{5a^2}{\tan 54}$$

Appendix A: Algebra Review

SECTION A.1: Algebraic Expressions

1. Undefined terms, definitions, axioms or postulates, theorems

2. Algebra; geometry

3. **a.** Reflexive
   b. Transitive
   c. Substitution
   d. Symmetric

4. **a.** $2 = 2$
   b. If $2 = x$, then $x = 2$.
   c. If $x = 2$ and $2 = y$, then $x = y$.
   d. If $2 + 5 = 7$ and $7 + y = z$, then $2 + 5 + y = z$.

5. **a.** 12
   b. –2
   c. 2
   d. –12

6. **a.** 8
   b. –8
   c. –22
   d. 1

7. **a.** 35
   b. –35
   c. –35
   d. 35

8. **a.** –84
   b. 84
   c. –84
   d. 84

9. No; Commutative Axiom for Multiplication

10. **a.** Commutative Axiom for Multiplication
    b. Associative Axiom for Addition
    c. Commutative Axiom for Addition
    d. Associative Axiom for Multiplication

11. **a.** 9
    b. –9
    c. 8
    d. –8

12. $(-3) - 7$

13. **a.** –4
    b. –36
    c. 18
    d. $-\dfrac{1}{4}$

14. 5 feet divided by 10 spaces = $\dfrac{1}{2}$ ft per space. Or since 5 feet = 60 inches, 60 inches divided by 10 spaces = 6 inches per space.

15. –$60

16. $25(2) + 30(2) = 50 + 60 = \$110$

17. **a.** $30 + 35 = 65$
    b. $28 - 12 = 16$
    c. $\dfrac{7}{2} + \dfrac{11}{2} = \dfrac{18}{2} = 9$
    d. $8x$

18. **a.** $54 - 24 = 30$
    b. $3(4 + 8) = 12 + 24 = 36$
    c. $7y - 2y = (7 - 2)y = 5y$
    d. $(16 + 8)x = 24x$

19. **a.** $(6 + 4)\pi = 10\pi$
    b. $(8 + 3)\sqrt{2} = 11\sqrt{2}$
    c. $7y - 2y = (7 - 2)y = 5y$
    d. $(9 - 2)\sqrt{3} = 7\sqrt{3}$

20. **a.** $(1 + 2)\pi r^2 = 3\pi r^2$
    b. $(7 + 3)xy = 10xy$
    c. $7x^2 y + 3xy^2$
    d. $(1 + 1)x + y = 2x + y$

21. **a.** $2 + 12 = 14$

    b. $5 \cdot 4 = 20$

    c. $2 + 3 \cdot 4 = 2 + 12 = 14$

    d. $2 + 6^2 = 2 + 36 = 38$

22. **a.** $9 + 16 = 25$

    b. $7^2 = 49$

    c. $9 + 6 \div 3 = 9 + 2 = 11$

    d. $[9 + 6] \div 3 = 15 \div 3 = 5$

23. **a.** $\dfrac{6}{-6} = -1$

    b. $\dfrac{8 - 6}{6 \cdot 3} = \dfrac{2}{18} = \dfrac{1}{9}$

    c. $\dfrac{10 - 18}{9} = \dfrac{-8}{9}$

    d. $\dfrac{5 - 12 + (-3)}{4 + 16} = \dfrac{-10}{20} = \dfrac{-1}{2}$

24. **a.** $8 + 10 + 12 + 15 = 45$

    b. $42 + 7 - 12 - 2 = 35$

25. **a.** $15 - 6 - 5 + 2 = 6$

    b. $12x^2 - 15x + 8x - 10 = 12x^2 - 7x - 10$

26. **a.** $10x^2 - 35x + 6x - 21 = 10x^2 - 39x - 21$

    b. $6x^2 - 10xy + 3xy - 5y^2 = 6x^2 - 7xy - 5y^2$

27. $5x + 2y$

28. $2xy + 2yz + 2xz$

29. $10x + 5y$

30. $xy + xz + y^2 + yz$; the total of the areas of the four smaller plots is also $xy + y^2 + xz + yz$.

31. $10x$

32. $9\pi + 48\pi + 9\pi = 66\pi$

SECTION A.2: Formulas and Equations

1. $5x + 8$

2. $-1x + 8$

3. $2x - 2$

4. $4x - 2$

5. $2x + 2 + 3x + 6 = 5x + 8$

6. $6x + 15 - 6x + 2 = 17$

7. $x^2 + 4x + 3x + 12 = x^2 + 7x + 12$

8. $x^2 - 7x - 5x + 35 = x^2 - 12x + 35$

9. $6x^2 - 4x + 5x - 10 = 6x^2 + 11x - 10$

10. $6x^2 + 9x + 14x + 21 = 6x^2 + 23x + 21$

11. $(a + b)(a + b) + (a - b)(a - b)$
$= a^2 + ab + ab + b^2 + a^2 - ab - ab + b^2$
$= 2a^2 + 2b^2$

12. $(x + 2)(x + 2) - (x - 2)(x - 2)$
$= x^2 + 2x + 2x + 4 - \left(x^2 - 2x - 2x + 4\right)$
$= x^2 + 4x + 4 - \left(x^2 - 4x + 4\right)$
$= x^2 + 4x + 4 - x^2 + 4x - 4$
$= 8x$

13. $4 \cdot 3 \cdot 5 = 60$

14. $5^2 + 7^2 = 35 + 49 = 74$

15. $2 \cdot 13 + 2 \cdot 7 = 26 + 14 = 40$

16. $6 \cdot 16 \div 4 = 96 \div 4 = 24$

17. $S = 2 \cdot 6 \cdot 4 + 2 \cdot 4 \cdot 5 + 2 \cdot 6 \cdot 5$
$S = 48 + 40 + 60$
$S = 148$

18. $A = \left(\dfrac{1}{2}\right) 2(6 + 8 + 10)$
$A = 1(24)$
$A = 24$

19. $V = \left(\dfrac{1}{3}\right)\pi(3)^2 \cdot 4$
$V = \dfrac{1}{3} \cdot \pi \cdot 9 \cdot 4$
$V = 12\pi$

20. $S = 4\pi r^2$
$S = 4\pi r(2)^2$
$S = 4\pi \cdot 4$
$S = 16\pi$

21. $2x = 14$
$x = 7$

22. $3x = -3$
$x = -1$

23. $\dfrac{y}{-3} = 4$
$y = -12$

24. $7y = -21$
$y = -3$

25. $2a + 2 = 26$
$2a = 24$
$a = 12$

26. $\dfrac{3b}{2} = 27$
$3b = 54$
$b = 18$

27. $2x + 2 = 30 - 6x + 12$
$2x + 2 = 42 - 6x$
$8x = 40$
$x = 5$

28. $2x + 2 + 3x + 6 = 22 + 40 - 4x$
$5x + 8 = 62 - 4x$
$9x = 54$
$x = 6$

29. Multiply equation by 6 to get
$2x - 3x = -30$
$-x = -30$
$x = 30$

30. Multiply equation by 12 to get
$6x + 4x + 3x = 312$
$13x = 312$
$x = 24$

31. Multiply equation by n to get
$360 + 135n = 180n$
$360 = 45n$
$8 = n$

32. Multiply equation by n to get
$(n - 2)180 = 150n$
$180n - 360 = 150n$
$-360 = -30n$
$12 = n$

33. $148 = 2 \cdot 5 \cdot w + 2 \cdot w \cdot 6 + 2 \cdot 5 \cdot 6$
$148 = 10w + 12w + 60$
$88 = 22w$
$4 = w$

34. $156 = \left(\dfrac{1}{2}\right) \cdot 12 \cdot (b + 11)$
$156 = 6(b + 11)$
$156 = 6b + 66$
$90 = 6b$
$15 = b$

35. $23 = \left(\dfrac{1}{2}\right)(78 - y)$
$46 = 78 - y$
$-32 = -y$
$32 = y$

36. $\dfrac{-3}{2} = \dfrac{Y - 1}{2 - (-2)}$
$\dfrac{-3}{2} = \dfrac{Y - 1}{4}$
$2Y - 2 = -12$
$2Y = -10$
$Y = -5$

SECTION A.3: Inequalities

1. The length of $\overline{AB}$ is greater than the length of $\overline{CD}$.

2. $e < f;\ f > e$

3. The measure of angle ABC is greater than the measure of angle DEF.

4. $x = 6,\ x = 9,\ x = 12$

5. **a.** $p = 4$

   b. $p = 10$

6. Yes

7. $AB > IJ$

8. The measure of angle JKL is greater than the measure of angle ABC.

9. **a.** False

   b. True

   c. True

   d. False

10. **a.** True

   b. False

   c. True

   d. False

11. The measure of the second angle must be greater than 148° and less than 180°.

12. The length of the second board must be less than 5 feet.

13. **a.** $-12 \le 20$

    b. $-10 \le -2$

    c. $18 \ge -30$

    d. $3 \ge -5$

14. **a.** $2 > -1$

    b. $12 < 18$

    c. $-12 > -18$

    d. $2 < 3$

15.

No Change	No Change
No Change	No Change
No Change	CHANGE
No Change	CHANGE

16. $5x \le 30$
$x \le 6$

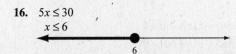

17. $2x \le 14$
$x \le 7$

18. $4x > 20$
$x > 5$

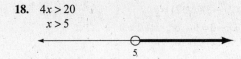

19. $-4x > 20$
$x < -5$

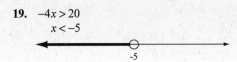

20. $10 - 5x \le 30$
$-5x \le 20$
$x \ge -4$

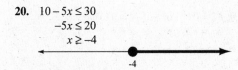

21. $5x < 200 - 5x$
$10x < 200$
$x < 20$

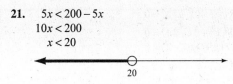

22. $5x + 10 < 54 - 6x$
$11x < 44$
$x < 4$

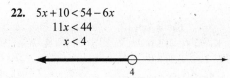

23. $2x - 3x \le 24$
$-x \le 24$
$x \ge -24$

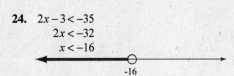

24. $2x - 3 < -35$
$2x < -32$
$x < -16$

25. $x^2 + 4x \le x^2 - 5x - 18$
$9x \le -18$
$x \le -2$

26. $x^2 + 2x < 2x - x^2 + 2x^2$
$x^2 + 2x < x^2 + 2x$
No solution or $\varnothing$.

27. Not true if $c < 0$.

28. Not true if $c = 0$.

29. Not true if $a = -3$ and $b = -2$.

30. Not true if $a = c$.

SECTION A.4: Quadratic Equations

1. **a.** 3.61

   b. 2.83

   c. −5.39

   d. 0.77

2. **a.** 4.12

   b. 20

   c. −2.65

   d. 1.26

3. a, c, d, f

4. a, b, c, e

5. **a.** $\sqrt{8} = \sqrt{4 \cdot 2} = 2\sqrt{2}$

   b. $\sqrt{45} = \sqrt{9 \cdot 5} = 3\sqrt{5}$

   c. $\sqrt{900} = 30$

   d. $\left(\sqrt{3}\right)^2 = 3$

6. **a.** $\sqrt{28} = \sqrt{4 \cdot 7} = 2\sqrt{7}$

   b. $\sqrt{32} = \sqrt{16 \cdot 2} = 4\sqrt{2}$

   c. $\sqrt{54} = \sqrt{9 \cdot 6} = 3\sqrt{6}$

   d. $\sqrt{200} = \sqrt{100 \cdot 2} = 10\sqrt{2}$

7. **a.** $\sqrt{\dfrac{9}{16}} = \dfrac{\sqrt{9}}{\sqrt{16}} = \dfrac{3}{4}$

   b. $\sqrt{\dfrac{25}{49}} = \dfrac{\sqrt{25}}{\sqrt{49}} = \dfrac{5}{7}$

   c. $\sqrt{\dfrac{7}{16}} = \dfrac{\sqrt{7}}{\sqrt{16}} = \dfrac{\sqrt{7}}{4}$

   d. $\sqrt{\dfrac{6}{9}} = \dfrac{\sqrt{6}}{\sqrt{9}} = \dfrac{\sqrt{6}}{3}$

8. **a.** $\sqrt{\dfrac{1}{4}} = \dfrac{\sqrt{1}}{\sqrt{4}} = \dfrac{1}{2}$

   b. $\sqrt{\dfrac{16}{9}} = \dfrac{\sqrt{16}}{\sqrt{9}} = \dfrac{4}{3}$

   c. $\sqrt{\dfrac{5}{36}} = \dfrac{\sqrt{5}}{\sqrt{36}} = \dfrac{\sqrt{5}}{6}$

   d. $\sqrt{\dfrac{3}{16}} = \dfrac{\sqrt{3}}{\sqrt{16}} = \dfrac{\sqrt{3}}{4}$

9. **a.** $\sqrt{54} \approx 7.35$ and $3\sqrt{6} \approx 7.35$

   b. $\sqrt{\dfrac{5}{16}} \approx 0.56$ and $\dfrac{\sqrt{5}}{4} \approx 0.56$

10. **a.** $\sqrt{48} \approx 6.93$ and $4\sqrt{3} \approx 6.93$

   b. $\sqrt{\dfrac{7}{9}} \approx 0.88$ and $\dfrac{\sqrt{7}}{3} \approx 0.88$

11.
$$x^2 - 6x + 8 = 0$$
$$(x - 4)(x - 2) = 0$$
$$x - 4 = 0 \quad \text{or} \quad x - 2 = 0$$
$$x = 4 \quad \text{or} \quad x = 2$$

12.
$$x^2 + 4x = 21$$
$$x^2 + 4x - 21 = 0$$
$$(x + 7)(x - 3) = 0$$
$$x + 7 = 0 \quad \text{or} \quad x - 3 = 0$$
$$x = -7 \quad \text{or} \quad x = 3$$

13.
$$3x^2 - 51x + 180 = 0$$
$$3\left(x^2 - 17x + 60\right) = 0$$
$$3(x - 12)(x - 5) = 0$$
$$x - 12 = 0 \quad \text{or} \quad x - 5 = 0$$
$$x = 12 \quad \text{or} \quad x = 5$$

14.
$$2x^2 + x - 6 = 0$$
$$(2x - 3)(x + 2) = 0$$
$$2x - 3 = 0 \quad \text{or} \quad x + 2 = 0$$
$$2x = 3 \quad \text{or} \quad x = -2$$
$$x = \frac{3}{2} \quad \text{or} \quad x = -2$$

15.
$$3x^2 = 10x + 8$$
$$3x^2 - 10x - 8 = 0$$
$$(3x + 2)(x - 4) = 0$$
$$3x + 2 = 0 \quad \text{or} \quad x - 4 = 0$$
$$3x = -2 \quad \text{or} \quad x = 4$$
$$x = -\frac{2}{3} \quad \text{or} \quad x = 4$$

16.
$$8x^2 + 40x - 112 = 0$$
$$8\left(x^2 + 5x - 14\right) = 0$$
$$8(x + 7)(x - 2) = 0$$
$$x + 7 = 0 \quad \text{or} \quad x - 2 = 0$$
$$x = -7 \quad \text{or} \quad x = 2$$

17.
$$6x^2 = 5x - 1$$
$$6x^2 - 5x + 1 = 0$$
$$(3x - 1)(2x - 1) = 0$$
$$3x - 1 = 0 \quad \text{or} \quad 2x - 1 = 0$$
$$3x = 1 \quad \text{or} \quad 2x = 1$$
$$x = \frac{1}{3} \quad \text{or} \quad x = \frac{1}{2}$$

18.
$$12x^2 + 10x = 12$$
$$12x^2 + 10x - 12 = 0$$
$$2(6x^2 + 5x - 6) = 0$$
$$2(3x - 2)(2x + 3) = 0$$
$$3x - 2 = 0 \quad \text{or} \quad 2x + 3 = 0$$
$$3x = 2 \quad \text{or} \quad 2x = -3$$
$$x = \frac{2}{3} \quad \text{or} \quad x = -\frac{3}{2}$$

19.
$$x^2 - 7x + 10 = 0$$
$$a = 1, \quad b = -7, \quad c = 10$$
$$x = \frac{-b \pm \sqrt{b^2 - 4ac}}{2a}$$
$$x = \frac{7 \pm \sqrt{49 - 4(1)(10)}}{2(1)}$$
$$x = \frac{7 \pm \sqrt{49 - 40}}{2}$$
$$x = \frac{7 \pm \sqrt{9}}{2}$$
$$x = \frac{7 + 3}{2} \text{ or } x = \frac{7 - 3}{2}$$
$$x = 5 \text{ or } 2$$

20.
$$x^2 + 7x + 12 = 0$$
$$a = 1, \quad b = 7, \quad c = 12$$
$$x = \frac{-b \pm \sqrt{b^2 - 4ac}}{2a}$$
$$x = \frac{-7 \pm \sqrt{49 - 4(1)(12)}}{2(1)}$$
$$x = \frac{-7 \pm \sqrt{49 - 48}}{2}$$
$$x = \frac{-7 \pm \sqrt{1}}{2}$$
$$x = \frac{-7 + 1}{2} \text{ or } x = \frac{-7 - 1}{2}$$
$$x = -3 \text{ or } -4$$

21.
$$x^2 + 9 = 7x$$
$$x^2 - 7x + 9 = 0$$
$$a = 1, \quad b = -7, \quad c = 9$$
$$x = \frac{-b \pm \sqrt{b^2 - 4ac}}{2a}$$
$$x = \frac{7 \pm \sqrt{49 - 4(1)(9)}}{2(1)}$$
$$x = \frac{7 \pm \sqrt{49 - 36}}{2}$$
$$x = \frac{7 \pm \sqrt{13}}{2} \approx 5.30 \text{ or } 1.70$$

22.
$$2x^2 + 3x = 6$$
$$2x^2 + 3x - 6 = 0$$
$$a = 2, \quad b = 3, \quad c = -6$$
$$x = \frac{-b \pm \sqrt{b^2 - 4ac}}{2a}$$
$$x = \frac{-3 \pm \sqrt{9 - 4(2)(-6)}}{2(2)}$$
$$x = \frac{-3 \pm \sqrt{9 + 48}}{4}$$
$$x = \frac{-3 \pm \sqrt{57}}{4} \approx 1.14 \text{ or } -2.64$$

23.
$$x^2 - 4x - 8 = 0$$
$$a = 1, \quad b = -4, \quad c = -8$$
$$x = \frac{-b \pm \sqrt{b^2 - 4ac}}{2a}$$
$$x = \frac{4 \pm \sqrt{16 - 4(1)(-8)}}{2(1)}$$
$$x = \frac{4 \pm \sqrt{16 + 32}}{2}$$
$$x = \frac{4 \pm \sqrt{48}}{2}$$
$$x = \frac{4 \pm 4\sqrt{3}}{2} \approx 5.46 \text{ or } -1.46$$

24.
$$x^2 - 6x - 2 = 0$$
$$a = 1, \quad b = -6, \quad c = -2$$
$$x = \frac{-b \pm \sqrt{b^2 - 4ac}}{2a}$$
$$x = \frac{6 \pm \sqrt{36 - 4(1)(-2)}}{2(1)}$$
$$x = \frac{6 \pm \sqrt{36 + 8}}{2}$$
$$x = \frac{6 \pm \sqrt{44}}{2}$$
$$x = \frac{6 \pm 2\sqrt{11}}{2}$$
$$x = 3 \pm \sqrt{11} \approx 6.32 \text{ or } -0.32$$

25.
$$5x^2 = 3x + 7$$
$$5x^2 - 3x - 7 = 0$$
$$a = 5, \quad b = -3, \quad c = -7$$
$$x = \frac{-b \pm \sqrt{b^2 - 4ac}}{2a}$$
$$x = \frac{3 \pm \sqrt{9 - 4(5)(-7)}}{2(5)}$$
$$x = \frac{3 \pm \sqrt{9 + 140}}{10}$$
$$x = \frac{3 \pm \sqrt{149}}{10} \approx 1.52 \text{ or } -0.92$$

26. $$2x^2 = 8x - 1$$
$$2x^2 - 8x + 1 = 0$$
$$a = 2, \ b = -8, \ c = 1$$
$$x = \frac{-b \pm \sqrt{b^2 - 4ac}}{2a}$$
$$x = \frac{8 \pm \sqrt{64 - 4(2)(1)}}{2(2)}$$
$$x = \frac{8 \pm \sqrt{64 - 8}}{4}$$
$$x = \frac{8 \pm \sqrt{56}}{4}$$
$$x = \frac{8 \pm 2\sqrt{14}}{4}$$
$$x = \frac{4 \pm \sqrt{14}}{2} \approx 3.87 \text{ or } 0.13$$

27. $$2x^2 = 14$$
$$x^2 = 7$$
$$x = \pm\sqrt{7}$$
$$x \approx \pm 2.65$$

28. $$2x^2 = 14x$$
$$2x^2 - 14x = 0$$
$$2x(x - 7) = 0$$
$$2x = 0 \quad \text{or} \quad x - 7 = 0$$
$$x = 0 \quad \text{or} \qquad x = 7$$

29. $$4x^2 - 25 = 0$$
$$4x^2 = 25$$
$$x^2 = \frac{25}{4}$$
$$x = \pm\frac{5}{2}$$

30. $$4x^2 - 25x = 0$$
$$x(4x - 25) = 0$$
$$x = 0 \quad \text{or} \quad 4x - 25 = 0$$
$$x = 0 \quad \text{or} \qquad 4x = 25$$
$$x = 0 \quad \text{or} \qquad x = \frac{25}{4}$$

31. $$ax^2 - bx = 0$$
$$x(ax - b) = 0$$
$$x = 0 \quad \text{or} \quad ax - b = 0$$
$$x = 0 \quad \text{or} \qquad ax = b$$
$$x = 0 \quad \text{or} \qquad x = \frac{b}{a}$$

32. $$ax^2 - b = 0$$
$$ax^2 = b$$
$$x^2 = \frac{b}{a}$$
$$x = \pm\sqrt{\frac{b}{a}}$$
$$x = \pm\frac{\sqrt{ab}}{a}$$

33. Let the length $= x + 3$ and width $= x$. The area is then:
$$x(x + 3) = 40$$
$$x^2 + 3x = 40$$
$$x^2 + 3x - 40 = 0$$
$$(x + 8)(x - 5) = 0$$
$$x + 8 = 0 \quad \text{or} \quad x - 5 = 0$$
$$x = -8 \quad \text{or} \qquad x = 5$$

Reject $x = -8$ because the length cannot be negative. The rectangle is 5 by 8.

34. $$x \cdot (x + 5) = (x + 1) \cdot 4$$
$$x^2 + 5x = 4x + 4$$
$$x^2 + 1x - 4 = 0$$
$$a = 1, \ b = 1, \ c = -4$$
$$x = \frac{-b \pm \sqrt{b^2 - 4ac}}{2a}$$
$$x = \frac{-1 \pm \sqrt{1 - 4(1)(-4)}}{2(1)}$$
$$x = \frac{-1 \pm \sqrt{1 + 16}}{2}$$
$$x = \frac{-1 \pm \sqrt{17}}{2}$$
$$CP = \frac{-1 + \sqrt{17}}{2} \approx 1.56$$

$\dfrac{-1 - \sqrt{17}}{2}$ is rejected because it is a negative number.

35. $$D = \frac{n(n - 3)}{2}$$
$$9 = \frac{n(n - 3)}{2}$$
$$18 = n^2 - 3n$$
$$0 = n^2 - 3n - 18$$
$$0 = (n - 6)(n - 3)$$
$$n = 6 \text{ or } n = -3$$
$$n = 6 \text{ reject } n = -3.$$

36. $D = \dfrac{n(n-3)}{2}$

$\quad n = \dfrac{n(n-3)}{2}$

$\quad 2n = n^2 - 3n$

$\quad 0 = n^2 - 5n$

$\quad 0 = n(n-5)$

$\quad n = 0 \quad \text{or} \quad n - 5 = 0$

$\quad n = 0 \quad \text{or} \quad\quad n = 5$

$\quad n = 5 \quad \text{reject } n = 0.$

37. $c^2 = a^2 + b^2$

$\quad c^2 = 3^2 + 4^2$

$\quad c^2 = 9 + 16$

$\quad c^2 = 25$

$\quad c = \pm 5$

$\quad c = 5; \text{ reject } c = -5$

38. $\quad\quad c^2 = a^2 + b^2$

$\quad\quad 10^2 = 6^2 + b^2$

$\quad 100 - 36 = b^2$

$\quad\quad\quad 64 = b^2$

$\quad\quad\quad\quad b = \pm 8$

$\quad\quad\quad\quad b = 8; \text{ reject } b = -8$